Elementary
Statistics

Elementary Statistics

Third Edition

Paul G. Hoel

Professor of Mathematics
University of California, Los Angeles

John Wiley & Sons, Inc.
New York London Sydney Toronto

Preface to the Third Edition

This book is designed for a one semester course for the student whose background in mathematics is limited to high school algebra. A number of mathematics departments offer a service course of this kind; however, in many schools a department with strong interests in the applications of statistics gives the course.

The present edition differs from the preceding one principally in organization and exposition. The basic material has been simplified slightly by eliminating some of the less important refinements and by presenting additional illustrations of the basic theory. The Special Topics material has been changed considerably by deleting the chapter on time series and index numbers and replacing it with a chapter on statistical decision methods. This was done because the former topics belong in a text designed exclusively for business and economics students and a new book *Basic Statistics for Business and Economics* by Hoel and Jessen contains that material.

Instructors who have used earlier editions of this book will discover a number of other modifications. Thus, expectation is introduced rather early and used in later chapters for developing theory, the correlation and regression chapters are amalgamated into one chapter, and the illustrations and exercises at the end of each chapter are labeled with the number of the section to which they belong. Several new exercises have been added and the old exercises have been modified so that the old answers are no longer correct, thus making dormitory solutions obsolete.

The last section of each chapter contains additional illustrations of the methods presented in that chapter. It may be used by the student as a basis for reviewing the entire chapter or as a source of additional illustrations for the individual sections of the chapter. To facilitate this type of usage, the illustrations have been labeled with the proper section number.

The material in the first nine chapters, exclusive of the sections marked with a boldface arrowhead, should suffice for the ordinary one semester course. The

v

arrowhead sections contain material for amplifying the basic course if time permits. The last four chapters have been included as additional optional material to enable the instructor with particular interests to include one or more topics related to those interests.

A number of individuals who have used earlier editions of this book were kind enough to offer suggestions for its improvement. Even though I was compelled to discard many of their proposals because of the necessity of compromising conflicting proposals, I very much appreciate their assistance.

Instructors can obtain answers to the even numbered problems in pamphlet form by writing to the publisher.

Los Angeles, California   PAUL G. HOEL
April 1971

Contents

CHAPTER 4
Probability Distributions

CHAPTER 5
Some Special Probability Distributions

CHAPTER 6
Sampling

CHAPTER 7
Estimation

CHAPTER 8
Testing Hypotheses

CHAPTER 9
Correlation and Regression

▶Special Topics

CHAPTER 10
Statistical Decision

CHAPTER 11
The Chi-Square Distribution

CHAPTER 12
Analysis of Variance

CHAPTER 13
Nonparametric Tests

CHAPTER 1

The Nature of Statistical Methods

1. INTRODUCTION

From a very general point of view, statistical methods are methods for treating numerical data. Thus, methods for collecting and analyzing the data of any business or government enterprise may be considered as belonging to the field of statistics. Such a definition, however, is much too broad in scope to be useful. It is necessary to restrict both the nature of the data and the reasons for studying them before such methods can rightfully be called statistical.

Statistical methods are concerned with data that have been obtained from taking observations, in the form of measurements or counts, from a source of such observations. For example, in studying the cost of medical services in a city, a small percentage of the inhabitants of the city would be selected and asked about their medical costs. Or, in studying public opinion on a controversial issue being debated in congress, a set of voters would be selected from across the country and asked their views concerning the issue.

Statisticians take observations of the type just described for the purpose of

1

drawing conclusions about the source of the observations. Thus, the medical costs of the selected individuals would be used to determine the general level of medical costs throughout the city. Similarly, the purpose of questioning only a small percentage of the voters on a controversial issue is to determine to a satisfactory approximation the opinions of all the voters on that issue.

The set of observations that is taken from some source of observations for the purpose of obtaining information about the source is called a *sample,* whereas the source of those observations is called a *population.* In view of the preceding discussion, *statistical methods may be described as methods for drawing conclusions about populations by means of samples.* The single word "statistics" is often used in place of statistical methods. Thus a student who is taking a course in statistics is taking a course in statistical methods.

At first glance, the foregoing definition may seem to be rather technical and contrary to the popular notion about statistics. For example, many business people look upon statistical methods as methods for collecting and summarizing business facts. The Federal Government employs a number of statisticians whose principal duty is to design efficient ways of collecting and summarizing various kinds of information. According to the preceding definition of statistical methods, these statisticians do not appear to be using statistical methods because they do not apply the information they have collected for drawing conclusions about the sources of the information. This viewpoint, however, does not take into account the fact that such information is gathered for the consumption of others who will use it to reach conclusions. Business concerns do not collect and summarize business facts just to admire the information obtained. They expect to use the information to make decisions, and whether or not they openly arrive at conclusions concerning the sources of the information the fact remains that they do make decisions on the basis of samples.

That part of statistical methods concerned with the collecting and summarizing of data is usually called *descriptive statistics.* The part concerned with drawing conclusions about the source of the data is called *statistical inference.* Since the ultimate objective is to make inferences, that is, draw conclusions, the descriptive part of statistics should be looked upon as a sort of preliminary to the main bout.

The use of statistical methods has increased remarkably in the last few decades, particularly in the biological and social sciences. Such methods have also proved very useful in various branches of the physical sciences and engineering. Because of this varied and strong interest, these methods have developed rapidly and have increased in complexity and diversity; nevertheless, many of the most important techniques are quite simple and are the same for all branches of application. Some of these universal methods are studied in this book. They should prove useful both to students who wish to understand how simple experimental data are handled and to students who need this type of background for more advanced work.

2.   ILLUSTRATIONS

This section describes a few problems of the type that statistical methods were designed to solve. It does not begin to cover the broad class of problems capable of being solved by statistical methods but rather illustrates a few of the simpler ones that can be solved by using only the methods developed in this book. One problem is of academic interest, whereas the others are typical real-life problems.

(a) The mayor of a large city wishes to know what his chances are for winning the primary nomination of his party to a vacancy in Congress. To obtain such information an organization engaged in promoting political campaigns agrees to take a poll of the voters of the state to determine the mayor's popularity. By using statistical methods, such an organization can decide how large a poll will be necessary in order to estimate, within any desired degree of accuracy, the percentage of voters who would favor the mayor over other leading candidates.

(b) A medical research team has developed a new serum it hopes will help to prevent a common children's disease. It wishes to test the serum. In order to assist the researchers in carrying out such a test, a school system in a large city has agreed to inoculate half of the children in certain grades with the serum. Records of all children in those grades are kept during the following year. On the basis of the percentages of those children who contract the disease during that year, both for the inoculated group and for the remaining half, it is possible to determine by statistical methods whether the serum is really beneficial.

(c) An industrial firm is concerned about the large number of accidents occurring in its plant. In the process of trying to discover the various causes of such accidents, an investigator considers factors related to the time of day. He collects information on the number of accidents occurring during the various working hours of the day, and by using statistical methods he is able to show that the accident rate actually increases during the morning and also during the afternoon. Further statistical studies then reveal some of the major contributing factors involved in those accidents.

One might be tempted to say that statistical methods are not needed in a problem such as this, and that all one needs to do is to calculate percentages and look at them to decide what is happening. If one has a large amount of properly selected data, such decisions will often be correct; however the high cost of collecting data usually forces one to work with only small amounts and it is precisely in such situations that statistical methods are needed to yield valid conclusions.

(d) A merchandizing firm wishes to determine the size of an advertising budget that will maximize the profit for one of its products. It decides to pick out three widely separated market areas and carry out promotional programs of varying intensity and costs that it believes are appropriate for that product. In doing so, it will need statistical methods on optimization to plan the study and estimate the best size budget to use.

(e) An instructor of an elementary statistics course is having difficulty convincing some of his students that the chances of winning from a slot machine are just as good immediately after someone has won some money as after a run of losses. For the purpose of convincing them, he and several of his students perform the following experiment on a slot machine in Las Vegas. The machine is played for one hour, or until the combined resources of instructor and students are exhausted, whichever occurs first. A record is kept of the number of wins and losses that occur immediately after a win, together with the amounts won, and also of the number of wins and losses, and amounts won, immediately after a run of, say, five losses. With data of this type available, the instructor should be able to apply statistical methods to convince the skeptics of his wisdom in this matter. Since a run of bad luck might make it difficult to demonstrate this wisdom, unless the machine were played a long while, the instructor would be well advised to come amply supplied with cash. An experiment of this type should also convince the students that slot machines are designed to extract money from naïve individuals.

3.   ESTIMATION AND HYPOTHESIS TESTING

An analysis of the preceding illustrations will show that they properly belong to the field of statistics because all are concerned with drawing conclusions about some population of objects or individuals and propose to do so on the basis of a sample from that population.

It may also be observed that these problems fall into two general categories. They are concerned either with estimating some property of the population or with testing some hypothesis about the population. The first illustration, for example, is concerned with estimating the percentage of the voters of the state who favor the mayor. The second illustration is one of testing the hypothesis that the percentage of children contracting a disease is the same for inoculated children as for children receiving no inoculation. The third illustration considers the problem of testing the hypothesis that the accident rate for a population of workers is constant over the day. The fourth illustration is concerned with estimating profit as a function of the amount of money spent on advertising the product. The fifth illustration is one of testing the hypothesis that the average amount of money won from a slot machine after a run of losses is the same as after a win.

Most of the statistical methods to be explained in this book are those for treating problems of these two types, namely, estimating properties of or testing hypotheses about populations. Although there are other types of conclusions or decisions that can be related to populations on the basis of samples, the bulk of those made by statisticians falls into one of the two aforementioned categories, and therefore they make up most of the material in this book.

4.   PROBABILITY

In the problem of estimating the percentage of voters who favor a candidate or issue, the solution will consist of a percentage based on the sample and a statement of the accuracy of the estimate, usually in a form such as "the probability is .95 that the estimate will be in error by less than 3 per cent." Similarly, in problems involving the testing of some hypothesis the decision to accept or reject the hypothesis will be based on certain probabilities.

It is necessary to use probability in such conclusions because a conclusion based on a sample involves incomplete information about the population, and therefore it cannot be made with certainty. The magnitude of the probability attached to a conclusion represents the degree of confidence one should have in the truth of the conclusion. The basic ideas and rules of probability are studied in a later chapter; meanwhile it should be treated from an intuitive point of view. Thus the statement that the probability is .95 that an estimate will be in error by less than 3 per cent should be interpreted as meaning that about 95 per cent of such statements made by a statistician are valid and about 5 per cent are not. In the process of studying statistical methods one will soon discover that probability is the basic tool of those methods.

Probability is an exceedingly interesting subject, even for those who have little liking for mathematics or quantitative methods. Many people enjoy some of the events associated with probability, if not the study itself; otherwise, how can one account for the large number of people who love to gamble at horse races, lotteries, cards, etc.? It may well be that it is their lack of probability sense that encourages them to gamble as they do. In any case, probability is used consciously or unconsciously by everyone in making all sorts of decisions based on uncertainty, and any student who wishes to be well educated, or to behave rationally, should have some knowledge of probability.

5.   ORGANIZATION

The study of the statistical methods discussed in the preceding sections will proceed by first considering properties of samples and then properties of populations. As indicated in section 1, such studies constitute the descriptive part of statistics. It will then be possible to consider the two basic problems of statistical inference, namely the problems of estimation and hypothesis testing. This means that for any given type of problem the sample data will always be studied first before any attempt is made to introduce a theoretical population from which the sample might have come. In Chapter 2, a beginning is made in the study of properties of sample data, after which some basic theoretical populations are introduced. It is at this theoretical stage that probability will appear.

CHAPTER 2

The Description of Sample Data

Since the purpose of this chapter is to study properties of samples taken from populations, it would seem necessary to agree first on how samples are to be taken because the desired properties may well depend on the method employed. Suppose, for example, that a student newspaper reporter has been assigned the task of determining the percentage of students having part-time jobs. He might attempt to estimate this percentage by polling the first 100 students he encountered in front of the student union. This method of sampling, however, is not likely to give a valid estimate of the population percentage because students found lolling in front of the union are often the campus loafers and social butterflies, and they are seldom the working type. The reporter would undoubtedly do much better if he were to select 100 cards blindly from the student enrollment card file and poll the selected students.

The problem of how to select a sample from a population so that valid conclusions about the population can be drawn from the sample is quite complicated.

This problem is discussed rather extensively in Chapter 5. In that discussion a method of sampling called random sampling is advocated and justified. Anticipating that material somewhat, an introduction is given here to this type of sampling. In its simplest form, when a single individual is to be selected from a population of individuals, the sampling is said to be *random* if each member of the population has the same chance of being chosen. Techniques used in games of chance are often employed to obtain such a sample. For example, at a large social affair at which a grand door prize is to be given away, it is customary to place all the numbered ticket stubs in a large container and then have a blindfolded individual select one ticket from the container, after the tickets have been thoroughly mixed. If, say, three individuals are to be selected from a population, the sampling will be random if every possible group of three individuals from the population has the same chance of being chosen. The preceding device that was employed to select one individual could also be used to select three individuals. Experience with such devices has shown that each individual, or group of individuals, is selected approximately the same number of times as every other individual, or group, when the experiment is repeated a large number of times, and therefore that they do conform to the requirement of showing no favoritism. In selecting a sample of students from a student body, choosing each tenth card from the enrollment files would undoubtedly be satisfactory as a practical method of random selection, provided the information desired about an individual is unrelated to his alphabetical position in the card file.

Among their various desirable properties, random samples tend to represent in miniature the population from which they are taken. This implies, for example, that if the experiment of selecting a student at random from the student body were repeated a number of times and the percentage of students who worked part time was calculated for the entire sample, that percentage would normally tend to get increasingly close to the true percentage for the student body as the sampling continued. Other types of sampling in which personal judgment enters usually do not possess this property.

As an illustration of this property of random samples generally representing the population being sampled, a random sampling experiment was carried out for an artificially constructed population. Suppose a large population of individuals can be classified into three groups. For example, it might be on the basis of age, those less than 30 years of age, those between 30 and 50 inclusive, and those over 50. Suppose further that the proportions of individuals in those three groups are $\frac{3}{6}, \frac{2}{6}$, and $\frac{1}{6}$, respectively. Then random sampling from such a population can be simulated by sampling from a population of 6 playing cards consisting of, say, 3 aces, 2 twos, and 1 three, provided the sampling is done properly. After each drawing of a card, the number is recorded and the drawn card is returned to the set. The cards are then thoroughly mixed and another drawing is made. This repeated mixing and returning of the drawn card to the set insures that the population proportions remain the same and that no favoritism will occur. If the original population were very large, the removal of some individuals from it by sampling would have no appreciable effect on the population proportions either. The preceding experiment was carried out 700 times in sets of 100 each. The accumulated results were reduced to percentages, calculated to the nearest decimal only, after each additional sample of 100 had been obtained. These random sample percentages, which are shown in Table 1, should be compared with the population percentages of 50, $33\frac{1}{3}$, and $16\frac{2}{3}$, respectively, to see how well the samples represent the population being sampled.

TABLE 1

Sample Size	100	200	300	400	500	600	700
Aces	44	47	45.7	47.8	49.4	49.6	50.0
Twos	35	34	35.0	33.2	32.8	32.8	32.9
Threes	21	19	19.3	19.0	17.8	17.5	17.1

The manner in which these sample percentages converge to their corresponding population percentages are more easily visualized geometrically, as shown in Fig. 1. They appear to converge well to the population percentages as the sample size increases.

Hereafter, whenever a sample is to be taken it will be assumed that it will be obtained by a random sampling method, even though the word random is not used explicitly.

Now turn to the problem of studying properties of samples taken from populations. In this connection, although the word "population" would seem to refer to a group of human individuals, in statistics it refers to a group of individuals, human or not, or objects of any kind. In studying samples and populations it is often assumed that interest is centered on a single particular property of members of the population. Thus, one might be interested in the property of weight in a population of students at a university, or in the color of the eyes in a population of insects, or in the percentage of iron in a population of meteorites. Samples are taken of individuals in a population, but then the property of interest is measured or counted for those individuals. The property that is measured or counted will usually be denoted by the letter X. Thus, X might represent

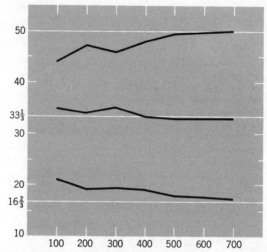

FIGURE 1. Random sampling results.

the weight of a college student, or the number of children in a family, or the annual medical costs of a medicare subscriber.

As an illustration of how one proceeds in the study of samples, consider the problem of what a physical education department at a university would do if it were interested in determining whether its male dormitory students are typical university students with respect to physical characteristics. In such a study it would undoubtedly wish to compare, as one source of information, the weight distribution of the dormitory students with that of nondormitory students. Now, weighing every male student on campus would certainly yield the desired information on weight distribution; however, this would become quite an undertaking in a large school at which such information is not required at registration time. The desired information, to sufficient accuracy, could be obtained much more easily by studying the weight distributions of samples of dormitory and non-dormitory students.

Suppose then that a random sample of, say, 120 students has been obtained from the dormitory population. Since the only concern here is what to do with samples, the nondormitory sample can be ignored in this discussion—it would be treated in the same manner as the dormitory sample. Suppose, furthermore, that the weights of these 120 students have been recorded to the nearest pound and that they range from the lightest at 110 pounds to the heaviest at 218 pounds.

It is very difficult to look at 120 measurements and obtain any reasonably accurate idea of how those measurements are distributed. For the purpose of obtaining a better idea of the weight distribution of the 120 students, it is therefore convenient to condense the data somewhat by classifying the measurements into groups. It will then be possible to graph the modified distribution and learn more about the original set of 120 measurements. This condensation is also useful for simplifying the computations of various averages that need to be evaluated, particularly if fast computing facilities are not available. These averages will supply additional information about the distribution. Thus the purpose of classifying data is to assist in the extraction of certain kinds of useful information from the data.

The weight measurements considered here comprise an example of observations made on what is called a *continuous variable*. This name is applied to variables, such as length, weight, temperature, and time, that can be thought of as capable of assuming any value in some interval of values. Thus the weight of a student in the 140–150 pound range can be deemed capable of assuming any value in this range. Variables such as the number of automobile accidents during a day, the number of beetles dying when they are sprayed with an insecticide, or the number of children in a family are examples of what is called a *discrete variable*. For the purposes of this book, discrete variables can be considered as variables whose possible values are integers; hence they involve counting rather than measuring.

Since any measuring device is of limited accuracy, measurements in real life are actually discrete in nature rather than continuous; however, this should not deter one from thinking about such variables as being continuous. Although the dormitory weights have been recorded to the nearest pound, they should be regarded as the values of a continuous variable, the values having been rounded off to the nearest integer. When a weight is recorded as, say, 152 pounds, it is assumed that the actual weight is somewhere between 151.5 and 152.5 pounds.

2.   CLASSIFICATION OF DATA

The problem of classifying the data of a sample usually arises only for continuous variables. As an illustration, consider the problem for the 120 dormitory weight measurements. What needs to be done is to choose a set of class intervals and then place each weight in its proper class. For example, if a 10-pound class interval is selected, all weights between, and including, 130 and 139 pounds would be placed in the same class. Since these weights were measured to the nearest pound,

TABLE 2

Class Boundaries	Frequencies
109.5–119.5	1
119.5–129.5	4
129.5–139.5	17
139.5–149.5	28
149.5–159.5	25
159.5–169.5	18
169.5–179.5	13
179.5–189.5	6
189.5–199.5	5
199.5–209.5	2
209.5–219.5	1

this particular class interval would extend from 129.5 to 139.5 pounds. The values 129.5 and 139.5 are called the *class boundaries* for that class. After the class boundaries have been determined it is a simple matter to list each measurement in its proper class interval. When this listing has been completed and the number of weights for each class interval recorded, the data are said to have been classified in a *frequency table*.

Table 2 illustrates the result of such a classification procedure for a set of 120 weights of the type under consideration.

3.   GRAPHICAL REPRESENTATION

Frequency distributions are easier to visualize if they are represented graphically. For continuous type variables a graph called a *histogram* is commonly used. It may also be used for discrete variables. The histogram for the frequency distribution of Table 2 is shown in Fig. 2. The class boundaries of Table 2 are marked off on the x axis starting and finishing at any convenient points. The frequency corresponding to any class interval is represented by the height of the rectangle whose base is the interval in question. The vertical axis is therefore the frequency axis. Histograms are particularly useful graphs for later work when frequency distributions of populations are introduced.

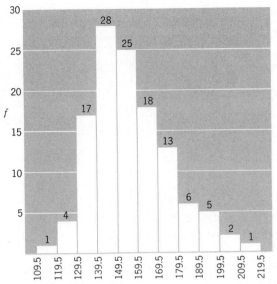

FIGURE 2. Distribution of the weights of 120 students.

The histogram of Fig. 2 is typical of many frequency distributions obtained from data found in nature and industry. They usually range from a rough bell-shaped distribution, such as that in Fig. 3, to something resembling the right half of a bell-shaped distribution, such as that in Fig. 5. A distribution of the latter

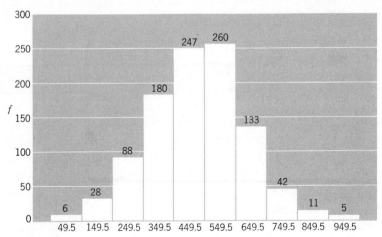

FIGURE 3. Distribution of 1,000 telephone conversations in seconds.

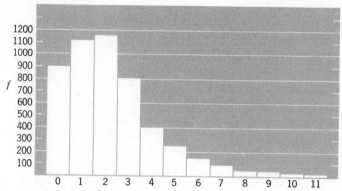

FIGURE 4.   Distribution of the number of children born to 5,000 women who have been married 20 years or more.

type is said to be *skewed*. Skewness refers to lack of symmetry with respect to a vertical axis. If a histogram has a long right tail and a short left tail, it is said to be skewed to the right; it is skewed to the left if the situation is reversed. The greater the unbalance, the greater the degree of skewness. It will be found, for example, that the following variables have frequency distributions that possess shapes of the type being discussed with approximately increasing degrees of skewness: stature, many industrial measurements, various linear biological measurements, weight, age at marriage, mortality age for certain diseases, and wealth. Figures 3, 2, 4, and 5 give the histograms for four typical distributions with increasing degrees of skewness.

4.   ARITHMETICAL DESCRIPTION

The principal reason for classifying data and drawing the histogram of the resulting frequency table is to determine the nature of the distribution. Some of the theory that is developed in later chapters requires that the distribution resemble the type of distribution displayed in Fig. 3; consequently, it is necessary to know whether one has this type of distribution before attempting to apply such theories to it.

Although a histogram yields a considerable amount of general information concerning the distribution of a set of sample measurements, more precise and useful information for studying a distribution can be obtained from an arithmet-

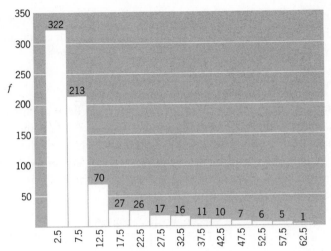

FIGURE 5.   Distribution of 727 deaths from scarlet fever classified according to age.

ical description of the distribution. For example, if the histogram of weights for a sample of nondormitory students were available for comparison with the histogram of the dormitory sample, it might be difficult to state, except in very general terms, how the two distributions differ. Rather than compare the two weight distributions in their entirety, it might suffice to compare the average weights and the variation in weights of the two groups. Such descriptive quantities are called arithmetic because they yield numbers, as contrasted to a histogram which is geometric in nature.

The nature of a statistical problem largely determines whether a few simple arithmetical properties of the distribution will suffice to describe it satisfactorily. Most of the problems that are encountered in this book are of the type that requires only a few properties of the distribution for its solution. For more advanced problems such a condensation of the information supplied by a distribution may not suffice. The situation here is similar to that arising when one discusses problems related to women's clothes and female beauty contests. A dress salesman, for example, might be satisfied to be told that a girl is 5 feet 7 inches tall and has measurements of 36–24–36 inches; however, a beauty contest judge would hardly be satisfied with such an arithmetical description. He most certainly would want to see the entire distribution.

Now there are many different kinds of quantities that can be used for an arithmetical description of a distribution. For most of the problems that are to be solved in this book it will suffice to have measures that describe where the histogram is located on the x axis and the extent to which the histogram is spread out along the x axis. Such measures are called measures of location and variation, respectively. They will be discussed in the next few sections.

4.1. The Mean

Suppose that a set of n sample values of a variable has been obtained from some population. These values are denoted by $X_1, X_2, \ldots, X_n$. This means that X_1 is the first sample value obtained, and X_n the last one. Thus, for the sample of 120 dormitory weights, X_1 would be the weight of the first student selected and X_{120} the weight of the last student selected. The familiar average of this set of numbers, which is denoted by $\overline{X}$, is given by the formula

$$(1) \qquad \overline{X} = \frac{X_1 + X_2 + \cdots + X_n}{n}.$$

Since there are other averages also used in statistics, it is necessary to give $\overline{X}$ a special name. It is called the arithmetic mean, or, for brevity, the *mean*.

Now consider the problem of calculating the mean when the data have been classified into a frequency table, such as the weight measurements in Table 2. For such data it is assumed that each of the original weights has been replaced by the weight at the midpoint of the class interval in which it lies. This midpoint value is called the *class mark* of the interval. Thus, all students whose weights fell in the 139.5–149.5 class interval would be assigned the weight 144.5 pounds, because that is the midpoint of the interval. From Table 2 it will be observed that there are 28 students whose weights would be listed as 144.5 pounds after the classification. As a result, for classified data the only values of X that arise in calculating the mean of a sample are the class mark values of the intervals. Suppose there are k class intervals in the frequency table. Let $x_1, x_2, \ldots, x_k$ denote the midpoint values of those k intervals and let $f_1, f_2, \ldots, f_k$ denote the corresponding frequencies in the frequency table.

To calculate the mean of the classified data by means of formula (1), it will suffice to add each classified value as many times as it occurs. For example, the value $x_4 = 144.5$ would be added 28 times in calculating the mean for the classified data of Table 2. In general, since x_1 occurs f_1 times it follows that x_1 must be added f_1 times, which is equivalent to multiplying x_1 by f_1. The same reasoning applies to the other class marks. The sum of all the measurements in a classified table is therefore the sum of all the products like $x_1 f_1$. The mean for such a table, which is denoted by $\overline{x}$, therefore assumes the form

$$(2) \qquad \overline{x} = \frac{x_1 f_1 + x_2 f_2 + \cdots + x_k f_k}{n}.$$

The preceding formulas can be written in much neater form if the summation symbol Σ (Greek sigma) is used. Since this symbol appears in other formulas also, it may be well to become acquainted with it now. It is merely a symbol which tells one to sum all the values of the quantity written after the symbol. Thus, formula (1) is written as

(3)
$$\overline{X} = \frac{\sum_{i=1}^{n} X_i}{n}.$$

The Σ symbol has n above it and $i = 1$ below it to indicate that all the numbers like X_i should be added, starting with $i = 1$ and finishing with $i = n$. The symbol X_i denotes the ith sample value, where i is some integer from 1 to n. In words, this symbol merely says, "sum all the values of the X's, beginning with the first ($i = 1$) sample value and ending with the last ($i = n$) sample value." In this condensed notation, formula (2) can be written in the form

(4)
$$\overline{x} = \frac{\sum_{i=1}^{k} x_i f_i}{n}.$$

Replacing each measurement of a sample by its class mark will usually yield a value of the mean that differs slightly from the mean of the original measurements; therefore the value of the mean given by formula (4) is only an approximation to the correct value given by formula (3). However, unless the classification is rather crude, the difference is usually so small that it can be ignored in most statistical problems.

Now it can be shown that the numerical value of the mean represents the point on the x axis at which a sheet of metal in the shape of the histogram of the distribution would balance on a knife edge. Since this balancing point is usually somewhere near the middle of the base of the histogram, it follows that $\overline{x}$ usually gives a fairly good idea where the histogram is located or centered. Thus the mean helps to describe a frequency distribution by telling where the histogram of the distribution is located along the x axis. It is therefore often called a *measure of location*.

For the frequency distribution of Table 2, whose histogram is shown in Fig. 2, calculations will show that the mean is approximately equal to 156. An inspection of Fig. 2 certainly shows that the histogram there ought to balance on a knife edge somewhere in the vicinity of 156. Thus the value of $\overline{x}$ here gives one a good idea of where the histogram of the distribution is located, or centered, on the x axis.

The principal reason for classifying a set of measurements is to obtain information concerning the nature of the distribution of the variable X being studied. Therefore, even though the data have been classified, the mean should be calculated by means of formula (3) rather than formula (4), provided the original set of measurements is still available.

Several other measures of location will be discussed in section 5.

4.2. The Standard Deviation

The purpose of this section is to introduce a quantity that measures the extent to which a set of measurements varies about the mean of the set. In this connection consider the two sets of measurements:

$$4, 5, 6, 7, 8 \quad \text{and} \quad 2, 4, 6, 8, 10$$

and their histograms shown in Fig. 6.

The mean of each set is 6. Since the common difference between consecutive pairs of measurements in the second set is twice what it is in the first set (2 to 1) and since the range of values of the second set is twice the range of values of the first set (8 to 4), most people would agree that the second set of measurements varies twice as much about its mean as does the first set.

The simplest measure of the variation of a set of measurements is undoubtedly the *range* of the set, that is, the difference between the largest and smallest measurement of the set. However, it has certain undesirable properties that will be discussed in section 5; therefore another more useful measure will be introduced.

Since the mean has been chosen as the desired measure of location of a set of measurements, a measure of variation should measure the extent to which the measurements deviate from their mean. For unclassified data corresponding to formula (3), the deviations are $X_1 - \overline{X}, X_2 - \overline{X}, \ldots, X_n - \overline{X}$. Values of X that are larger than $\overline{X}$ will produce positive deviations, whereas values of X smaller

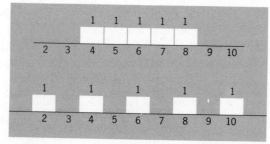

FIGURE 6.   Two artificial distributions.

than $\overline{X}$ will yield negative deviations. Since only positive distances from the mean are desired, it is necessary to use the absolute values of such deviations, or possibly use their squares. Since absolute values are difficult to work with, it is customary to take the squares of the deviations and average them. Thus, a measure of variation of a set of measurements about the mean of the set is given by

(5)
$$\frac{\sum\limits_{i=1}^{n} (X_i - \overline{X})^2}{n}.$$

If a set of measurements has been classified and formula (4) is being used as the measure of location, then the average of the squares of the deviations is

(6)
$$\frac{\sum\limits_{i=1}^{k} (x_i - \overline{x})^2 f_i}{n}.$$

In the solutions of statistical inference problems in later chapters, a slight modification of this measure is more useful than the measure itself. This modification consists in using the divisor $n - 1$ in place of n in these formulas. The justification for this modification will be given in Chapter 5. The resulting quantity is denoted by the special symbol s^2 and is called the *sample variance*. Thus, by definition, for unclassified data

(7)
$$s^2 = \frac{\sum\limits_{i=1}^{n} (X_i - \overline{X})^2}{n - 1}$$

and for classified data

(8)
$$s^2 = \frac{\sum\limits_{i=1}^{k} (x_i - \overline{x})^2 f_i}{n - 1}.$$

Since the variance involves the squares of deviations, it is a number in squared units. In many problems it is desirable that quantities describing a distribution possess the same units as the original set of measurements. The mean satisfies this requirement but the variance does not; however, by taking the positive square root of the variance the desired effect can be achieved. The resulting quantity is called the standard deviation. Thus, by definition, the *standard deviation* for unclassified data is

(9)
$$s = \sqrt{\frac{\sum\limits_{i=1}^{n} (X_i - \overline{X})^2}{n - 1}}$$

and for classified data is

(10)
$$s = \sqrt{\frac{\sum\limits_{i=1}^{k}(x_i - \bar{x})^2 f_i}{n - 1}}.$$

Just as in the case of calculating the mean, one should use formula (9) rather than (10) in calculating the standard deviation, provided the original unclassified data are available. However, unless the classification is very coarse, the difference between the values of s obtained from formulas (9) and (10) will be very small and can be ignored in practice.

It is sometimes more convenient to calculate s^2 for classified data by means of the following formula rather than by means of the definition:

(11)
$$s^2 = \frac{1}{n - 1}\left[\sum_{i=1}^{k} x_i^2 f_i - n\bar{x}^2\right].$$

This formula is easily derived from (8) by squaring the binomial term and summing the individual terms. A corresponding formula for unclassified data can also be written down, if desired, by analogy with the preceding formula by deleting f_i and replacing x by X.

4.3. Interpretation of the Standard Deviation

For the purpose of observing how s measures variation, consider once more the two sets of measurements corresponding to Fig. 6. Since $\bar{X} = 6$ for both sets, calculation of s for the two sets by means of formula (9) will yield the values

$$s_1 = \sqrt{\frac{4 + 1 + 0 + 1 + 4}{4}} = \frac{\sqrt{10}}{2}$$

and

$$s_2 = \sqrt{\frac{16 + 4 + 0 + 4 + 16}{4}} = \sqrt{10}.$$

The fact that the value of the standard deviation is twice as large for the second set as for the first set is certainly a satisfying property of the standard deviation if it is to be considered a measure of variation.

The preceding illustration, which indicates that the standard deviation increases in size as the variability of the data increases, does not give any clue to the meaning of the magnitude of the standard deviation. Thus, if the values of the standard deviation in this illustration had been $3\sqrt{10}$ and $6\sqrt{10}$, instead of $\sqrt{10}/2$

and $\sqrt{10}$, the interpretation would have been the same. This situation is similar to that occurring when two students compare scores on a test. One student may score twice as many points as the other, but this does not reveal how much absolute knowledge of the subject either student possesses.

In order to give some quantitative meaning to the size of the standard deviation, it is necessary to anticipate certain results of later work. For a set of data that has been obtained by sampling a particular type of population, called a normal population, it will be shown later that when the sample is large the interval from $\bar{x} - s$ to $\bar{x} + s$ usually includes about 68 per cent of the observations and that the interval from $\bar{x} - 2s$ to $\bar{x} + 2s$ usually includes about 95 per cent of the observations. A sample from a population of this type usually has a histogram that looks somewhat like the histogram of Fig. 3. As the sample increases in size, the histogram tends to approach the shape of a bell.

As an illustrative example of this property, consider the data of Table 2. Computations using formulas (4) and (10) yielded the values $\bar{x} = 156.2$ and $s = 19.1$. The two intervals $(\bar{x} - s, \bar{x} + s)$ and $(\bar{x} - 2s, \bar{x} + 2s)$ therefore become (137.1, 175.3) and (118.0, 194.4), respectively. These values have been marked off with vertical lines on the x axis of Fig. 7, which is the graph of the histogram for the frequency distribution of Table 2.

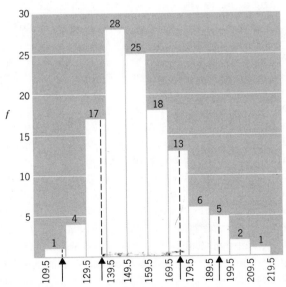

FIGURE 7. Histogram for the distribution of 120 weights.

In computing the percentages of the data lying within each of these two intervals, it is necessary to approximate how many observations in an interval lie to the right, and to the left, of a point inside the interval. Toward this end, it is assumed that all the observations in an interval are spread uniformly along the interval. Thus, for the second interval, it is assumed that one of the four observations lies in the first quarter of the interval, another is in the second quarter of the interval, etc. With this understanding, the number of observations lying between 137.1 and 175.3 is approximately equal to $4 + 28 + 25 + 18 + 8 = 83$. The values of 4 and 8 are obtained by interpolating according to the preceding agreement on how observations are spread out. For example, 4 is obtained by realizing that the distance from 129.5 to 137.1 is 7.6 units, whereas the distance from 129.5 to 139.5 is 10 units; therefore 76 per cent of the 17 observations in this interval should be assumed to be to the left of 137.1 and 24 per cent to the right. Since 24 per cent of 17 is equal to 4, to the nearest integer, it follows that 4 of the 17 observations of this interval should be assumed to lie to the right of 137.1. The value of 8 in the preceding sum is obtained in a similar manner. Thus the distance from 169.5 to 175.3 is 5.8 units, whereas the distance from 169.5 to 179.5 is 10 units; hence 58 per cent of the 13 measurements in this interval should be treated as being to the left of the point 175.3. Since 58 per cent of 13 is 7.54, or 8 to the nearest integer, it follows that 8 of those 13 observations should be assumed to be to the left of 175.3. Now since the total number of observations is 120 and since $83/120 = .69$, it follows that approximately 69 per cent of the observations fall inside the interval $(\bar{x} - s, \bar{x} + s)$. Similar calculations will show that approximately 94 per cent of the observations fall inside the interval $(\bar{x} - 2s, \bar{x} + 2s)$. These results are certainly close to the theoretical percentages of 68 and 95 for normal distributions.

For a distribution whose histogram resembles the histogram in Fig. 4 or Fig. 5, one would not expect to find the percentages for the intervals discussed to be very close to the theoretical percentages for a normal distribution, yet the percentages are often fairly close to those theoretical values.

By using the foregoing geometrical interpretation of the standard deviation, it is possible to obtain a rough idea of the size of the standard deviation for familiar distributions. Consider, for example, the distribution of stature for adult males. One might guess that about 95 per cent of all adult males would have heights somewhere between 5 feet 2 inches and 6 feet 2 inches. Since this is a 12-inch interval, and since for a normal distribution 95 per cent of the observations would be expected to lie in the interval $(\bar{x} - 2s, \bar{x} + 2s)$ whose length is $4s$, one would guess that $4s = 12$ inches, or that $s = 3$ inches. A crude estimate of the standard deviation of the frequency distribution of stature for adult males is therefore 3 inches. In a similar manner, one should be able to give a crude estimate of the size of the standard deviation for other familiar frequency distributions and thus acquire a feeling for the standard deviation as a measure of the variability of data.

▶ 5.   OTHER DESCRIPTIVE MEASURES

It is interesting to observe in newspaper reports how different groups will employ different averages to describe the distribution of wages in their industry. Employers usually quote the mean wage to indicate the economic status of employees. Labor leaders, however, prefer to use the mode or the median as an indicator of the wage level.

The *mode* of a set of measurements is defined as the measurement with the maximum frequency, if there is one. Thus, for the set of measurements 3, 3, 4, 4, 4, 5, 5, 6, 6, 7, 8, 9, 9, the mode is 4. In certain industries there may be more laborers working at the lowest wage scale than at any of the other scales, and therefore labor leaders would naturally prefer to quote the mode in describing the distribution of wages. In most problems of this type one must know more than just a measure of location, and therefore the mode is likely to be of limited value here.

The *median* of a set of measurements is defined as the middle measurement, if there is one, after the measurements have been arranged in order of magnitude. For the set of measurements in the preceding paragraph, which is arranged in order of magnitude, the median is 5. If there is an even number of measurements, one chooses the median to be halfway between the two middle measurements. Thus, if one of the 5's were deleted from the preceding set, there would be no middle measurement and the median would become 5.5. The median is a more realistic measure to describe the wage level in certain industries than either the mode or the mean. Since the median wage is one such that half the employees receive at least this much and half receive at most this much, one usually obtains a fairly good picture of the wage level from the median. The mean has the disadvantage that if most of the wages are fairly low, but there is a small percentage of very high wages, the mean wage will indicate a deceptively high wage level. The median would seem to be better than the mean here as an indicator of what is popularly meant by the wage level.

The median has another rather attractive property. If the variable being studied is income and incomes are listed in intervals of 500 dollars, but with all over

20,000 dollars listed as 20,000 dollars or more, it is not possible to compute the mean income because of the uncertainty of the incomes in the last interval. The median income would be unaffected by this lack of information and therefore could substitute for the mean here.

In view of the foregoing remarks about some of the attractive properties of the median, one might wonder why statisticians usually prefer the mean to the median. There are some computational advantages of the mean over the median; however, the principal reason for preferring the mean is that it is a much more useful and reliable measure to use in making statistical inferences. Since the ultimate objective is to solve statistical inference problems, the mean is usually preferred to the median when both measures can be found.

Several measures of variation, in addition to the standard deviation, are occasionally used. The *range,* which is such a measure, has already been defined in section 4.2. It is a popular measure in such fields as industrial quality control and meteorology. This popularity rests principally on its ease of computation and interpretation. It is a simple matter to find the largest and smallest measurements of a set, in contrast to calculating the standard deviation of the set. It is also a simple matter to explain how the range measures variation, in contrast to explaining how the standard deviation does so. However, an unfortunate property of the range is the tendency of its value to increase in size as the sample size increases. For example, one would expect the range in the weights of a sample of ten college students to be considerably smaller than the range in the weights of a sample of 100 students. It is possible to adjust the range for this growth by means of fancy formulas, but then the range loses its simplicity.

Measures of variation can also be constructed by methods similar to that used to define the median. Instead of merely finding a value of x that divides the histogram into a lower and upper half, one could also find two other values of x, one dividing the histogram at a point such that one-fourth of the area is to the left of it and the other such that one-fourth of the area is to the right of it. These two measures, together with the median, constitute the three *quartiles* of the distribution. The smallest quartile is called the first quartile, the median is called the second quartile, and the largest quartile is called the third quartile. A simple measure of variation can be constructed from the quartiles by taking the difference between the third quartile and the first quartile. This measure, which is used in some fields, is called the *interquartile range.*

One could construct other measures of variation in a similar manner by considering deciles rather than quartiles. Deciles are values of x that divide the histogram into tenths. Thus one might use the difference between the ninth decile and the first decile as a measure of variation. Measures of this general type possess certain advantages over the standard deviation similar to those possessed by the median over the mean. However, the same kind of superiority of the standard deviation exists here as exists for the mean over the median when it comes to solving problems of statistical inference.

These ideas can be extended further by introducing percentiles, which are values of x that divide the histogram into 100 equal area parts. Percentiles are used extensively in such fields as psychology and educational testing, for example in comparing students' results on standardized examinations with national averages. Undoubtedly some of you had the experience of being told at what percentile you rated on a scholastic aptitude test.

6.	ADDITIONAL ILLUSTRATIONS

This section is designed to serve as a problem solving review of the material presented in this chapter, or, if preferred, as additional illustrations of the problem solving techniques introduced in the various sections of the chapter. The section number is attached to each part of a problem that possesses several parts to facilitate the additional illustration usage of these problems.

1. The following classified data represent the number of hours worked per week for 100 laborers.

x	11	14	17	20	23	26	29	32	35	38	41	44	47	50	53
f	2	1	2	1	6	7	11	12	12	17	19	9	0	0	1

Using these data (a) 3 draw the histogram, (b) 4.1 calculate the mean, (c) 4.2 calculate the standard deviation, (d) 4.4 calculate the percentage of the measurements lying inside the two intervals $(\bar{x} - s, \bar{x} + s)$ and $(\bar{x} - 2s, \bar{x} + 2s)$, (e) 5 estimate the median by inspecting the histogram, (f) 5 estimate the first and third quartiles and the interquartile range by inspecting the histogram. The solutions follow.

(a)

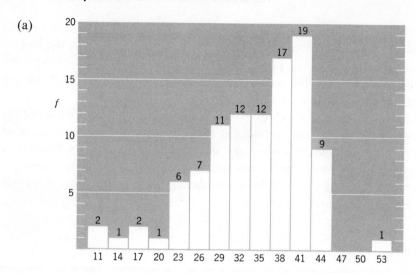

(b) and (c)

x_i	f_i	$x_i f_i$	x_i^2	$x_i^2 f_i$
11	2	22	121	242
14	1	14	196	196
17	2	34	289	578
20	1	20	400	400
23	6	138	529	3,174
26	7	182	676	4,732
29	11	319	841	9,251
32	12	384	1,024	12,288
35	12	420	1,225	14,700
38	17	646	1,444	24,548
41	19	779	1,681	31,939
44	9	396	1,936	17,424
47	0			
50	0			
53	1	53	2,809	2,809
	100	3,407		122,281

Using formula (4),

$$\bar{x} = \frac{3407}{100} = 34.07 = 34.1.$$

Using formula (11),

$$s^2 = \frac{1}{99}[122{,}281 - 116{,}068] = \frac{6213}{99} = 62.8.$$

Hence, $s = 7.9$.

(d) $(\bar{x} - s, \bar{x} + s) = (26.2, 42.0)$

$(\bar{x} - 2s, \bar{x} + 2s) = (18.3, 49.9)$

Measurements inside the first interval total

$$\frac{1.3}{3}(7) + 11 + 12 + 12 + 17 + \frac{2.5}{3}(19) = 71; \text{ hence 71 per cent.}$$

Measurements outside the second interval total

$$2 + 1 + \frac{2.8}{3}(2) + 1 = 6; \text{ hence 94 per cent inside.}$$

(e) Median $= 35.5$

(f) First quartile $= 29.1$

Third quartile $= 40.1$

Interquartile range $= 40.1 - 29.1 = 11$

2. The following data give the weekly wages of 100 laborers of a foreign country. (a) 2 Classify the data into a frequency table. Choose a class interval of length 2 and begin $\frac{1}{2}$ unit below the smallest measurement. Thus, choose boundaries 38.5–40.5, 40.5–42.5, etc., and class marks 39.5, 41.5, etc. (b) 3 Draw the histogram. (c) 4.1 Calculate the mean. (d) 4.2 Calculate the standard deviation. (e) 4.4 Calculate the percentage of measurements lying inside the interval $(\bar{x} - s,$ $\bar{x} + s)$. (f) 5 Estimate the median. (g) 4.1 Calculate the mean of the unclassified data and compare with the answer of part (c). (h) 4.2 Calculate the standard deviation of the unclassified data and compare with the answer of part (d).

49 47 51 48 50 46 53 46 45 50 49 50 50 47 56 51 46 47 54 53 48 50 51 50 60

51 46 48 52 52 46 61 52 49 50 45 57 54 51 60 50 56 52 44 49 45 51 50 40 46

54 47 50 55 55 47 48 53 50 49 45 50 50 51 47 54 43 53 55 50 53 52 52 51 47

51 48 45 44 50 52 49 51 51 47 53 49 46 61 49 52 48 39 46 52 51 57 49 45 50

(a)

x	39.5	41.5	43.5	45.5	47.5	49.5	51.5	53.5	55.5	57.5	59.5	61.5
f	2	0	3	14	14	25	21	10	5	2	2	2

(b)

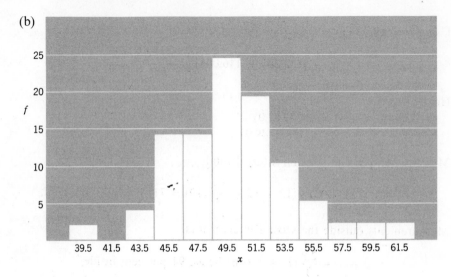

(c) $\bar{x} = 50.0$
(d) $s = 4.0$
(e) 71 per cent
(f) 50
(g) $\bar{x} = 50.0$; hence the same to this accuracy.
(h) $s = 4.0$; hence the same to this accuracy.

EXERCISES

Section 1

1. Toss a coin 500 times, recording the number of heads obtained in sets of 50. Calculate the cumulative proportion of heads obtained in decimal form and observe whether the resulting set of 10 proportions appears to approach $\frac{1}{2}$.

2. Roll a die 500 times, recording only the number of times that a 1 or a 6 appears in sets of 50 rolls. Calculate the cumulative proportion of such results and observe whether the proportions seem to approach $\frac{1}{3}$.

Section 3

3. Given the following frequency table of the diameters in feet of 56 shrubs from a common species, draw its histogram showing the class marks.

x	1	2	3	4	5	6	7	8	9	10	11	12
f	1	7	12	16	10	5	4	2	1	1	0	1

4. Given the following frequency table of the heights in centimeters of 1,000 students, draw its histogram showing the class marks.

x	155–157	158–160 etc.												
f	4	10	26	51	90	145	188	180	130	90	58	22	4	1	1

5. The following data are for the traveling time to and from work in hours per day for a group of aircraft workers. Draw the histogram, assuming continuous time.

Under 1 hour	80
1 up to 2	42
2 up to 3	7
3 up to 4	4
4 up to 5	3
5 up to 6	2

6. Given the following frequency distribution of the number of children born to wives aged 40–44 during their married lives, draw its histogram.

x	0	1	2	3	4	5	6	7	8	9	10 or more
f	1230	1520	1545	962	537	301	174	108	69	51	73

7. What type of distribution would you expect grade point averages of college students to possess? Sketch your idea of the nature of this distribution with the proper units on the x axis.

8. If you were to study the age distribution of college students, would you consider age to be a discrete or continuous variable? What would you choose for class boundaries?

9. Draw a sample of 100 one-digit random numbers from Table II in the appendix by taking consecutive digits from as many columns as needed. Classify these numbers into a frequency table and draw the histogram, even though the variable here is discrete. These numbers, which are discussed in Chapter 5, possess a "rectangular" distribution; hence your histogram should resemble a rectangle.

Section 4.1

10. For the histogram of problem 3, guess the value of $\bar{x}$. Do the same for the histograms of problems 4 and 5.

11. For the data of problem 3, calculate $\bar{x}$.

12. For the data of problem 4, calculate $\bar{x}$.

Section 4.2

13. Without classifying the data, calculate (a) the mean, (b) the standard deviation for the following set of weights of 11 children: 39, 51, 39, 45, 46, 53, 48, 40, 43, 47, 44.

14. Without classifying the data, calculate (a) the mean, (b) the standard deviation for the following set of grade point averages of 20 students: 2.4, 1.2, 1.4, 2.4, 1.0, 1.8, 1.8, 1.4, 1.8, 3.2, 2.4, 2.2, 2.4, 1.8, 3.6, 1.8, 1.2, 2.4, 1.8, 3.4.

15. For the data of problem 3, calculate s.

16. For the data of problem 4, calculate s.

Section 4.3

17. For the histogram of problem 3, using the results in problems 11 and 15, calculate the approximate percentages of the data that lie within the intervals $\bar{x} \pm s$ and $\bar{x} \pm 2s$.

18. For the histogram of problem 4, using the results of problems 12 and 16, calculate the approximate percentages of the data that lie within the intervals $\bar{x} \pm s$ and $\bar{x} \pm 2s$.

19. Show that (a) $\dfrac{1}{n} \sum_{i=1}^{n} (x_i + c) = \bar{x} + c$, (b) $\sum_{i=1}^{n} (x_i - \bar{x}) = 0$.

20. Suppose you found that 5 per cent of the shots on a target were a radial distance of more than three standard deviations of such distances from the center. What might you conclude about the distribution of the shots?

21. If shoe sizes of college male (or female) students were assumed to possess a normal distribution, what would you guess the standard deviation of shoe size to be if, by your knowledge of shoe sizes, you estimated a two-standard deviation interval about the mean?

22. What can be said about a distribution if $s = 0$?

23. If the scores on a set of examination papers are changed by (a) adding 10 points to all scores, (b) increasing all scores by 10 per cent, what effect will these changes have on the mean and on the standard deviation?

24. Cite some type of data for which you feel the standard deviation would tend to exaggerate the amount of variation present in the data.

25. Calculate the standard deviation for the discrete distribution consisting of the two points $x = -2$ and $x = 2$ with frequencies of 50 each and comment about the 68 per cent and 95 per cent interpretation of s here.

Section 5

▶**26.** Cite some type of data for which you feel the median would be a more appropriate measure of location than the mean.

▶**27.** For the data of problem 13, calculate the median and the range.

▶**28.** For the data of problem 3, calculate the median and the range.

▶**29.** For the data of problem 3, calculate the interquartile range and the ninth decile.

Section 6

▶**30.** As a review exercise, use your results from problem 9 to work the problems that were solved in the first review exercise of section 6.

CHAPTER 3

Probability

As indicated in Chapter 1, the solutions to the statistical problems posed there are given in terms of probability statements. Although probability is applied to a variety of practical situations, an understanding of the subject is made much simpler if it is applied to nonpractical situations, such as those that arise in certain games of chance. For this reason, the definition and the rules of probability are presented in terms of idealized problems, but it is assumed that the same rules may later be applied to practical statistical problems.

Before discussing probability, it is necessary to discuss experiments that can be repeated or that can be conceived of as being repeatable. Tossing a coin, reading the daily temperature on a thermometer, or counting the number of bad eggs in a carton are examples of a simple repetitive experiment. An experiment in which several rabbits are fed different rations in an attempt to determine the relative growth properties of the rations may be performed only once with those same animals; nevertheless, the experiment may be thought of as the first in an

33

unlimited number of similar experiments, and therefore it may be considered repetitive. Selecting a sample from a population is a repetitive experiment and is, of course, the type of experiment that is of particular interest in solving statistical problems.

Consider a simple repetitive experiment such as tossing a coin twice or, what is equivalent, tossing two distinct coins simultaneously. In this experiment there are four possible outcomes of interest; they are denoted by

HH, HT, TH, TT.

The symbol HT, for example, means that a head is obtained on the first toss and a tail on the second toss. If the experiment had consisted of tossing the coin three times, there would have been eight possible outcomes of the experiment, which would be denoted by

HHH, HHT, HTH, THH, HTT, THT, TTH, TTT.

The three letters in a group here express the outcomes of the three tosses in the given order. An experiment such as reading the temperature on a thermometer, however, has an infinite number of possible outcomes, since the temperature is a continuous type of variable. In this discussion of probability only experiments with a finite number of possible outcomes are considered. Other types are discussed in Chapter 4.

For any experiment to which probability is to be applied, it is first necessary to decide what possible outcomes of the experiment are of interest, and to make a list of all such outcomes. This list must be such that when the experiment is performed, exactly one of the outcomes will occur. In the experiment of tossing a coin three times, interest was centered on whether the coin showed a head or a tail on each of the tosses; therefore all the possible outcomes are those that were listed previously. In selecting a digit from the table of random digits in Table II in the appendix one might be interested in knowing which digit was obtained, in which case there are ten possible outcomes corresponding to the digits 0, 1, ..., 9. However, one might be interested only in knowing whether the digit was less than 3 in magnitude. Then there would be only two possible outcomes of the experiment, namely obtaining a digit that is less than 3 or obtaining a digit that is at least as large as 3. A game of chance experiment that will be used frequently for illustrative purposes is the experiment of drawing one ball from a box of balls of different colors. Thus, suppose a box contains three red, two black, and one green ball. Then interest will be centered only on what color a drawn ball is and not on which particular ball is obtained. Here there are three possible outcomes of the experiment corresponding to the three colors. Another interesting game of chance experiment is the experiment of rolling two dice. If it is assumed that one can distinguish between the two dice and interest centers on what number of points shows on each of the dice, then there are 36 possible outcomes, because each die has six possible outcomes and these outcomes can be paired in all possible ways. Table 1 gives a list of the possible outcomes.

TABLE 1					
11	21	31	41	51	61
12	22	32	42	52	62
13	23	33	43	53	63
14	24	34	44	54	64
15	25	35	45	55	65
16	26	36	46	56	66

The first number of each pair denotes the number that came up on one of the dice, and the second number denotes the number that came up on the other. If the two dice are not distinguishable, it is necessary to roll them in order rather than simultaneously.

It is convenient, in developing the theory of probability, to visualize things geometrically and to represent each of the possible outcomes of an experiment by means of a point. Thus, in the experiment of tossing a coin three times one would use eight points. It makes no difference what points are chosen as long as one knows which point corresponds to which possible outcome. Each point is labeled with a letter or symbol to indicate the outcome that it represents. Since possible outcomes are often called simple events, the letter e with a subscript corresponding to the number of the outcome in the list of possible outcomes is customarily used in labeling a point. In the coin tossing experiment, for example, one could label the eight points by means of $e_1, e_2, \ldots, e_8$. Thus e_1 would represent the event of obtaining HHH and e_2 that of obtaining HHT. Since a label such as HHT is self explanatory there seems little point to introducing additional labels here. However, it is easier to write e_2 than HHT; consequently the e symbol does possess an advantage. It is convenient in developing the rules of probability to give a name to these basic sets of points. This is done formally as follows:

> **Definition.** *The set of points representing the possible outcomes of an experiment is called the sample space for the experiment.*

A sample space for the coin-tossing experiment is shown in Fig. 1, in which the numbering of the points is shown by the symbols directly above the points.

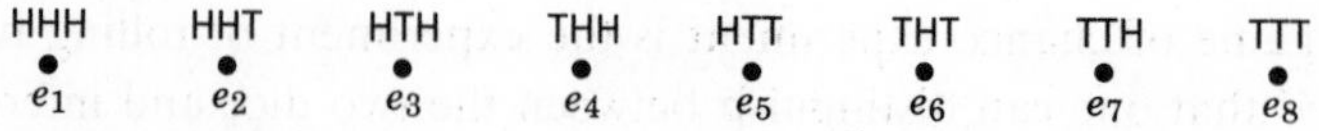

HHH	HHT	HTH	THH	HTT	THT	TTH	TTT
•	•	•	•	•	•	•	•
e_1	e_2	e_3	e_4	e_5	e_6	e_7	e_8

FIGURE 1.   A sample space for a coin-tossing experiment.

A natural sample space to choose for the experiment of selecting a digit from the table of random digits consists of the 10 points on the x axis corresponding to the integers $0, 1, \ldots, 9$. This is shown in Fig. 2. The letter e was not used in Fig. 2 because the labeling of a point by means of its x coordinate is about as simple as one could desire.

For the experiment of choosing a colored ball from the box of balls, the sample space consists of three points, which have been labeled in the order of the colors given and which are shown in Fig. 3.

A convenient sample space for the experiment of rolling two dice is a double array of 36 points, with six rows and six columns, attached to the elements of Table 1. This sample space will not be illustrated because Table 1 and your imagination should suffice. The symbols of Table 1 are highly descriptive of the experimental outcomes and are also very simple; there is no point in introducing a new set of symbols here.

Thus far there has been no apparent reason for introducing a geometrical representation of the outcomes of an experiment; however, later in the development of the theory the advantage of this approach will become apparent.

The next step in the construction of a mathematical model for an experiment is to attach numbers to the points in the sample space that will represent the relative frequencies with which those outcomes are expected to occur. If the experiment of tossing a coin three times were repeated a large number of times and a cumulative record kept of the proportion of those experiments that produced, say, three heads, one would expect that proportion to approach $\frac{1}{8}$ because each of the eight possible outcomes would be expected to occur about equally often. Actual experiments of this kind usually show that such expectations are justified, provided the coin is well balanced and is tossed vigorously. In view of such considerations, the number $\frac{1}{8}$ would be attached to each of the points in the sample space shown in Fig. 1. The number assigned to the point labeled e_i

FIGURE 2.   A sample space for a random-digit experiment.

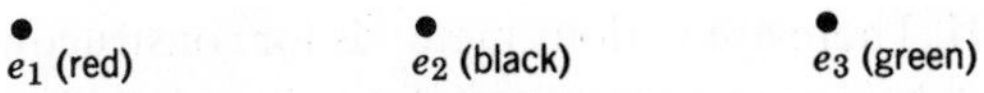

•
e_1 (red)

•
e_2 (black)

•
e_3 (green)

FIGURE 3. Sample space for a colored ball experiment.

in a sample space is called the probability of the event e_i and is denoted by the symbol $P\{e_i\}$. Thus, in the coin-tossing experiment each of the events $e_1, e_2, \ldots, e_8$ possesses the probability $\frac{1}{8}$.

If the experiment of selecting a digit from the table of random digits is carried out a large number of times, it will be found that each of the ten digits $0, 1, \ldots, 9$ will be obtained with approximately the same relative frequency, and therefore that the experimental relative frequency for each of the digits will be close to $\frac{1}{10}$. On the basis of such experience each of the sample points in the sample space shown in Fig. 2 would be assigned the probability $\frac{1}{10}$.

The situation for the experiment corresponding to the sample space shown in Fig. 3 is somewhat different from the preceding ones. It is no longer true that each of the possible outcomes would be expected to occur with the same relative frequency. If the balls are well mixed in the box before each drawing and the drawn ball is always returned to the box so that the composition of the box is unchanged, one would expect to obtain a black ball twice as often as a green ball and a red ball three times as often as a green ball. This implies that in repeated sampling experiments one would expect the relative frequencies for the three colors red, black, and green to be close to $\frac{3}{6}$, $\frac{2}{6}$, and $\frac{1}{6}$, respectively. Thus, the three points e_1, e_2, and e_3 in Fig. 3 would be assigned the probabilities $\frac{3}{6}$, $\frac{2}{6}$, and $\frac{1}{6}$, respectively.

The experiment of rolling two dice is treated in much the same manner as the coin-tossing experiment. Symmetry and experience suggest that each point in the sample space corresponding to Table 1 should be assigned the probability $\frac{1}{36}$.

The preceding experiments illustrate how one proceeds in general to assign probabilities to the points of a sample space. If the experiment is one for which symmetry and similar considerations suggest what relative frequencies are to be expected for the various outcomes, then those expected relative frequencies are chosen as the probabilities for the corresponding points. This was the basis for the assignment of probabilities in the coin-tossing experiment, the colored ball experiment, and the die-rolling experiment. If no such symmetry considerations are available but experience with the given experiment is available, then the relative frequencies obtained from such experience can be used for the probabilities to be assigned. The assignment of probabilities for the sample space of Fig. 2 was based partly on experience and partly on faith in the individuals who constructed Table II. There are various methods for constructing tables of random digits, some of them being very complicated. In all such tables it is to be expected that each digit will occur about the same number of times and that there will be no discernible patterns in sequences of digits. For example, the pair 12 should not occur more frequently than any other pair, say the pair 74. However, since such sets of digits are often based on physical devices that are assumed to produce digits possessing such properties, it is unreasonable to expect a set of such digits to behave in this ideal manner. A good approximation is all that can be hoped for.

Since the probabilities assigned to the points of a sample space are either the expected relative frequencies based on symmetry considerations or the long-run experimental relative frequencies, probabilities must be numbers between 0 and 1 and their sum must be 1, because the sum of a complete set of relative frequencies is always 1. In the experiments related to coin tossing, colored balls, and dice rolling, the probabilities obviously sum to 1 because they were constructed that way. If the probabilities for the random-digit experiment had been based entirely on the relative frequencies obtained in a long run of experiments, then those probabilities would obviously sum to 1 because the sum of all the relative frequencies must be 1.

Now, in any given experimental situation, whether academic or real, it is the privilege of the statistician to assign any probabilities he desires to the possible outcomes of the experiment, provided they are numbers between 0 and 1 and provided they sum to 1. Of course if he is sensible he will try to assign numbers that will represent what he believes or knows to be the long-run relative frequencies for those outcomes, otherwise his mathematical model is not likely to represent the actual experiment satisfactorily and therefore his conclusions derived on the basis of the model are likely to be erroneous.

It is usually quite easy to assign satisfactory probabilities to the possible outcomes of games of chance; however, this is not the case for most real-life experiments. For example, if the experiment consists of selecting an individual at random from the population of a city and interest is centered on whether the individual will die during the ensuing year, then there is no satisfactory way of assigning a probability here other than by using the experience of insurance companies. If one were interested in determining proper insurance premiums, it would be necessary to assign probabilities of death at the various ages. These are usually chosen to be the values obtained from extensive experience of insurance companies over the years. Since mortality rates have been decreasing over the years for most age groups, any mortality table based on past experience is likely to be out of date for predicting the future. Thus the probabilities assigned on the basis of past experience may not be very close to the actual relative frequencies existing today, and therefore the premiums calculated from them will not be very accurate. Fortunately for the insurance companies, premiums calculated on the basis of past experience are larger than they would be if they had been based on more up-to-date experience.

In view of the foregoing discussion, it follows that the probability of a simple event is to be interpreted as a theoretical, or idealized, relative frequency of the event. This does not imply that the observed relative frequency of the event will necessarily approach the probability of the event for an increasingly large number of experiments because one may have chosen an incorrect model; however, one hopes that it will. Thus, if one has a supposedly honest die, one would hope that the observed relative frequency of, say, a 4 showing would approach the probability $\frac{1}{6}$ as an increasingly large number of rolls is made; but one would not be too upset if it did not approach $\frac{1}{6}$ because of the imperfections in any manufactured article and because of the difficulty of simulating an ideal experiment. In this connection, it should be noted that the operators of gambling houses have done well financially by assuming that dice do behave as expected. They have certainly rolled dice enough to check on such assumptions. Of course, if experience shows that a die is not behaving as expected, they will replace it very quickly with a new die.

Constructing theoretical models to explain nature is the chief function of scientists. If the models are realistic, the conclusions derived from them are likely to be realistic. A probability model is such a model designed to enable one to draw conclusions about relative frequencies of experimental outcomes.

Although probability is being interpreted here in terms of expected relative frequency, it is also used by some individuals as a measure of their degree of belief in the future occurrence of an event, even when there is no conceptually repeatable experiment involved. Thus, they might be willing to assign a probability to the event that there will be no major war in the world during the next ten years. Such a probability should correspond to the betting odds that the individual will give that a war will not occur. Probabilities of this type, which are called personal, or subjective, probabilities are frequently employed in business problems and will be discussed further in this connection in Chapter 10.

2. PROBABILITY OF AN EVENT

Now that a geometrical model has been constructed for an experiment, consisting of a set of points with labels $e_1, e_2, \ldots$ to represent all the possible outcomes and a corresponding associated set of probabilities $P\{e_1\}, P\{e_2\}, \ldots$, the time has come to discuss the probability of composite events. The possible outcomes e_1, $e_2, \ldots$ of a sample space are called *simple events*. A *composite event* is defined as a collection of simple events. For example, the event of obtaining exactly two heads in the coin-tossing experiment of Fig. 1 is a composite event that consists of the three simple events e_2, e_3, and e_4. Similarly, the event of obtaining a random digit smaller than 4 in the random-digit experiment of Fig. 2 is a composite event consisting of the four simple events $x = 0, 1, 2, 3$. Composite events are usually denoted by capital letters such as A, or B, or C.

Now since the probabilities assigned to the simple events of Fig. 1, namely $\frac{1}{8}$, represent expected relative frequencies for their occurrence, one would expect the composite event consisting of the simple events e_2, e_3, and e_4 to occur in about three-eighths of such experiments in the long run of such experiments. Similarly, for the experiment represented by Fig. 2, one would expect the composite event consisting of the simple events $x = 0, 1, 2, 3$ to occur in the long run in about four-tenths of such experiments. In view of such expectations, and because probability has been introduced as an idealization of relative frequency, the following definition of probability for a composite event should seem very reasonable.

(1)   **Definition.**   *The probability that a composite event A will occur is the sum of the probabilities of the simple events of which it is composed.*

As an illustration, if A is the event of obtaining two heads in tossing a coin three times, it follows from this definition and the sample space in Fig. 1 that

$$P\{A\} = P\{e_2\} + P\{e_3\} + P\{e_4\} = \tfrac{3}{8}.$$

Similarly, if B is the event of getting a digit smaller than 4 in selecting a random digit, it follows from this definition and Fig. 2 that

$$P\{B\} = P\{0\} + P\{1\} + P\{2\} + P\{3\} = \tfrac{4}{10}.$$

As another illustration, let C be the event of getting a red or a green ball in the experiment for which Fig. 3 is the sample space. Since C is composed of the events e_1 and e_3, it follows that

$$P\{C\} = P\{e_1\} + P\{e_3\} = \tfrac{4}{6}.$$

As a final illustration for which the composite events are not quite so obvious, consider once more the experiment of rolling two dice. Table 1, with points associated with each outcome and with the probability $\frac{1}{36}$ attached to each point, can serve as the sample space here. First, let E be the event of getting a total of 7 points on the two dice. The simple events that yield a total of 7 points are the following: 16, 25, 34, 43, 52, 61. The sum of the corresponding six probabilities of $\frac{1}{36}$ each therefore gives $P\{E\} = \frac{6}{36} = \frac{1}{6}$. Next, let F be the event of getting a total number of points that is an even number. Simple events such as 11, 13, 22, etc. satisfy the requirement of yielding an even numbered total. The sum of two even digits, or the sum of two odd digits, will yield an even number. From Table 1 it will be observed that there are 18 points of this type; hence it follows that $P\{F\} = \frac{18}{36} = \frac{1}{2}$. Finally, let G be the event that both dice will show at least 4 points. Here the simple events such as 44, 45, 56, etc. will satisfy. Table 1 shows that there are 9 such; therefore $P\{G\} = \frac{9}{36} = \frac{1}{4}$.

In many games-of-chance experiments the various possible outcomes are expected to occur with the same relative frequency; therefore all the points of the sample space for such experiments are assigned the same probability, namely $1/n$, where n denotes the total number of points in the sample space. This was true, for example, in the experiments of coin tossing, random-digit selection, and dice rolling. It was not true, however, for the colored ball experiment. When the experiment is of this simple type, that is when all the simple-event probabilities are equal, the calculation of the probability of a composite event is very easy. It consists of merely adding the probability $1/n$ as many times as there are simple events comprising the composite event. Thus, if the composite event A consists of a total of $n(A)$ simple events, the value of $P\{A\}$ can be expressed by the simple formula

$$(2) \qquad P\{A\} = \frac{n(A)}{n}.$$

In the experiment of rolling two dice, for example, the probability of obtaining a total of 7 points is obtained by counting the number of points in the sample space given in Table 1 that produce a 7 total, of which there are 6, and dividing this number by the total number of points, namely 36.

Although it is not often possible in real-life problems to use formula (2), it is easier to work with than is the general definition (1) involving the addition of probabilities; therefore, it alone is used in the next few sections to derive basic formulas. The formulas obtained in this manner can be shown to hold equally well for the general definition and therefore are applicable to all types of problems. Since only the formulas are needed in applied problems, there will be no appreciable loss in the understanding of how to solve practical problems by following this procedure.

3.   ADDITION RULE

Applications of probability are often concerned with a number of related events rather than with just one event. For simplicity, consider two such events, A_1 and A_2, associated with an experiment. One may be interested in knowing whether both A_1 and A_2 will occur when the experiment is performed. This joint event is denoted by the symbol (A_1 and A_2) and its probability by $P\{A_1$ and $A_2\}$. On the other hand, one may be interested in knowing whether at least one of the events A_1 and A_2 will occur when the experiment is performed. This event is denoted by the symbol (A_1 or A_2) and its probability by $P\{A_1$ or $A_2\}$. At least one of the two events will occur if A_1 occurs but A_2 does not, if A_2 occurs but A_1 does not, or if both A_1 and A_2 occur. Thus, the word "or" here means "or" in the sense of either one, the other, or both. The purpose of this section is to obtain a formula for $P\{A_1$ or $A_2\}$.

If two events A_1 and A_2 possess the property that the occurrence of one prevents the occurrence of the other, they are called *mutually exclusive* events. For example, let A_1 be the event of getting a total of 7 in rolling two dice, and A_2 the event of getting a total of 11: then A_1 and A_2 are mutually exclusive events. For mutually exclusive events there are no outcomes that correspond to the occurrence of both A_1 and A_2; therefore the two events do not possess any points in common in the sample space. This is shown schematically in Fig. 4. In this diagram the points lying inside the two light regions labeled A_1 and A_2 represent the simple events that yield the composite events A_1 and A_2, respectively. If $n(A_1)$ denotes the

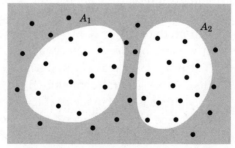

FIGURE 4. Sample space with two mutually exclusive events.

number of points lying inside the region labeled A_1 and $n(A_2)$ the number lying inside the region labeled A_2, then the total number of points associated with the occurrence of either A_1 or A_2 is the sum of those two numbers; consequently, if n denotes the total number of sample points, it follows from formula (2) that

$$P\{A_1 \text{ or } A_2\} = \frac{n(A_1) + n(A_2)}{n}$$

$$= \frac{n(A_1)}{n} + \frac{n(A_2)}{n}$$

Since the last two fractions are precisely those defining $P\{A_1\}$ and $P\{A_2\}$, this result yields the desired addition formula, which may be expressed as follows:

(3)   ***Addition Rule.***   *When A_1 and A_2 are mutually exclusive events,*

$$P\{A_1 \text{ or } A_2\} = P\{A_1\} + P\{A_2\}.$$

For more than two mutually exclusive events, it is merely necessary to apply this formula as many times as required. A slightly more complicated formula can be derived for events that are not mutually exclusive; however, there will be no occasion to use such a formula in later sections and so it is omitted here. It can be found in one of the exercises at the end of this chapter.

In the preceding illustration of rolling two dice, in which A_1 and A_2 denoted the events of getting a total of 7 and 11 points, respectively, the probability of getting either a total of 7 or a total of 11 can be obtained by means of formula (3). From Table 1 it is clear that A_1 and A_2 contain no points in common and, from counting points, that $P\{A_1\} = \frac{6}{36}$ and $P\{A_2\} = \frac{2}{36}$; therefore

$$P\{A_1 \text{ or } A_2\} = \frac{6}{36} + \frac{2}{36} = \frac{8}{36}.$$

This result is, of course, the same as that obtained by counting the total number of points, namely 8, that yield the composite event $(A_1 \text{ or } A_2)$ and applying formula (2) directly.

As another illustration, what is the probability of getting a total of at least 10 points in rolling two dice? Let A_1, A_2, and A_3 be the events of getting a total of exactly 10 points, 11 points, and 12 points, respectively. From Table 1 it is clear that these events have no points in common and that their probabilities are given by $P\{A_1\} = \frac{3}{36}$, $P\{A_2\} = \frac{2}{36}$, and $P\{A_3\} = \frac{1}{36}$. Therefore, by formula (3), the probability that at least one of those mutually exclusive events will occur is given by

$$P\{A_1 \text{ or } A_2 \text{ or } A_3\} = \frac{3}{36} + \frac{2}{36} + \frac{1}{36} = \frac{1}{6}.$$

This result also could have been obtained directly from Table 1 by counting favorable and total outcomes. Although the formula does not seem to possess any advantage here over direct counting for problems related to Table 1, it is a very useful formula for problems in which probabilities of events are available but for which tables of possible outcomes are not.

4.   MULTIPLICATION RULE

The purpose of this section is to obtain a formula for $P\{A_1 \text{ and } A_2\}$ in terms of probabilities of single events. In order to do so, it is necessary to introduce the notion of conditional probability. Suppose one is interested in knowing whether A_2 will occur subject to the condition that A_1 is known to have occurred or else is certain to occur. The geometry of this problem is shown in Fig. 5. It is assumed here that A_1 and A_2 are not mutually exclusive events.

Since A_1 must occur, the only experimental outcomes that need be considered are those corresponding to the occurrence of A_1. The sample space for this problem is therefore reduced to the simple events that comprise A_1. They are represented in Fig. 5 by the points lying inside the region labeled A_1. Among those points, the ones that also lie inside the region labeled A_2 correspond to the occurrence of both A_1 and A_2. They are the points that lie in the overlapping parts of A_1 and A_2. Let $n(A_1)$ denote the number of points lying inside A_1 and let $n(A_1 \text{ and } A_2)$ denote the number that lie inside both A_1 and A_2. Then, from formula (2), the probability that A_2 will occur if the sample space is restricted to be the set of points inside A_1 is given by the ratio $n(A_1 \text{ and } A_2)/n(A_1)$. But this probability

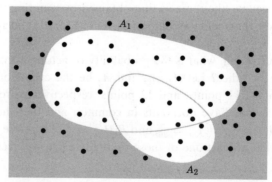

FIGURE 5.   A sample space for conditional probability.

is what is meant by the probability that A_2 will occur subject to the restriction that A_1 must occur. If this latter probability is denoted by the new symbol $P\{A_2 | A_1\}$, then

(4)
$$P\{A_2 | A_1\} = \frac{n(A_1 \text{ and } A_2)}{n(A_1)}.$$

As an illustration of the application of this formula, a calculation will be made of the probability that the sum of the points obtained in rolling two dice is 7, if it is known that the dice showed at least 3 points each. Let A_1 denote the event that two dice will show at least 3 points each and let A_2 denote the event that two dice will show a total of 7 points. The sample space for this problem is shown in Fig. 6; it was obtained directly from Table 1.

The points that comprise the event A_1 are all the points except those in the first two rows and the first two columns of Fig. 6. They are shown inside the rectangle of Fig. 6. Here $n(A_1) = 16$. The points that comprise the event A_2 are the diagonal points shown in Fig. 6. The number of points that lie inside A_2 which also lie inside A_1 is seen to be $n(A_1 \text{ and } A_2) = 2$. As a result, formula (4) gives the result

$$P\{A_2 | A_1\} = \frac{2}{16} = \frac{1}{8}.$$

What this means in an experimental sense is that in the repeated rolling of two dice one discards all those experimental outcomes in which either die showed a number of points less than 3. Then among the experimental outcomes that are retained one calculates the proportion of them that yielded a total of 7 points.

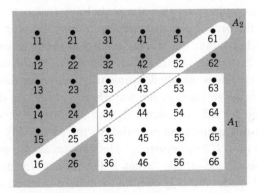

FIGURE 6. Sample space for a conditional probability problem.

This proportion in the long run of experiments should approach $\frac{1}{8}$. It is interesting to note that the chances of getting a total of 7 points is less when one knows that both dice show at least 3 points than under ordinary rolls.

Now consider formula (4) in general terms once more. From Fig. 5 and formula (2) it is clear that

$$P\{A_1\} = \frac{n(A_1)}{n}$$

and that

$$P\{A_1 \text{ and } A_2\} = \frac{n(A_1 \text{ and } A_2)}{n}.$$

Dividing the second of these two expressions by the first and canceling n will give

$$\frac{P\{A_1 \text{ and } A_2\}}{P\{A_1\}} = \frac{n(A_1 \text{ and } A_2)}{n(A_1)}.$$

This result in conjunction with (4) will yield the formula

(5) $$P\{A_2|A_1\} = \frac{P\{A_1 \text{ and } A_2\}}{P\{A_1\}}.$$

This formula defines the *conditional probability* of A_2 given A_1 and when written in product form yields the fundamental multiplication formula for probabilities, which may be expressed as follows:

(6)   **Multiplication Rule.** $P\{A_1 \text{ and } A_2\} = P\{A_1\}P\{A_2|A_1\}.$

In words, this formula states that the probability that both of two events will occur is equal to the probability that the first event will occur, multiplied by the conditional probability that the second event will occur when it is known that the first event is certain to occur. Either one of the two events may be called the first event, since this is merely convenient language for discussing them, and there is no time order implied in the way they occur. Even though there is no time order implied for the two events A_1 and A_2 in the symbol $P\{A_2|A_1\}$, it is customary to call this conditional probability "the probability that A_2 will occur when it is known that A_1 has occurred." Thus, if you are being dealt a five-card poker hand, someone might ask, "what is the probability that your hand will contain the ace of spades if it is known that you have received the ace of hearts?" This is merely convenient language for discussing probabilities of a poker hand that must contain the ace of hearts, and there is no implication that if the ace of spades is in the hand it was obtained after obtaining the ace of hearts. For many pairs of events, however, there is a definite time-order relationship. For example, if A_1 is the event that a high school graduate will go to college and A_2 is the event that he will graduate from college, then A_1 must precede A_2 in time.

As an illustration of the multiplication rule, a calculation will be made of the probability of getting two red balls in drawing two balls from a box containing three red, two black, and one green ball. It will be assumed here that the first ball drawn is not returned to the box before the second drawing is made. This experiment differs from the one that was considered in Fig. 3, since there repetitions of the experiment involved returning the drawn ball each time.

Let A_1 denote the event of getting a red ball on the first drawing and A_2 that of getting a red ball on the second drawing. In order to be able to continue using formula (2) rather than the more general definition (1), it is necessary to give each ball a number and use six points in the sample space as shown in Fig. 7. The first three will represent red balls, the next two the black balls, and the last one the green ball. Then, by formula (2),

$$P\{A_1\} = \frac{3}{6}.$$

For the purpose of calculating $P\{A_2|A_1\}$ it suffices to consider only those experimental outcomes for which A_1 has occurred. Since the first ball drawn is not returned to the box, this means considering only those experiments in which the first ball drawn was one of the three red balls. Thus, the second part of the experiment can be treated as a new single experiment in which one ball is to be drawn from a box containing two red, two black, and one green ball. As a result, $P\{A_2|A_1\}$ represents the probability of getting a red ball in drawing one ball from this new box of balls. Here there are five points in the sample space

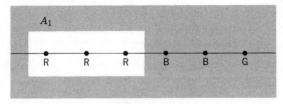

FIGURE 7. Sample space for first drawing.

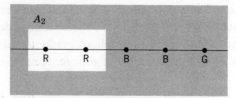

FIGURE 8. Sample space for
second drawing.

as shown in Fig. 8, the first two representing red balls, the next two black balls,
and the last one a green ball. Hence, by formula (2),

$$P\{A_2|A_1\} = \frac{2}{5}.$$

Application of formula (6) to these two results then gives

$$P\{A_1 \text{ and } A_2\} = \frac{3}{6} \cdot \frac{2}{5} = \frac{1}{5}.$$

The advantage of using formula (6) on this problem will become apparent if
one tries to solve this problem by applying formula (2) directly to the sample
space that corresponds to this two-stage experiment. That sample space will consist
of thirty points and will resemble the sample space shown in Fig. 6, except that
the main diagonal points will be missing because they correspond to getting the
same numbered colored ball on both drawings. This two-dimensional sample
space is shown in Fig. 9. Using Fig. 9 the solution is obtained by counting the

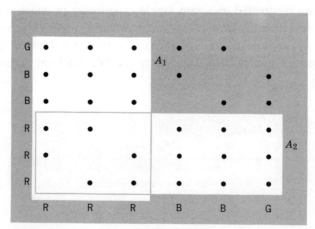

FIGURE 9. Sample space for a two-stage
experiment.

points common to A_1 and A_2 and dividing by the total number of points. This gives $\frac{6}{30} = \frac{1}{5}$. The advantage of formula (6) is most pronounced in two-stage and multiple-stage experiments, which usually possess complicated sample spaces with a large number of points, because it reduces the calculation of probabilities to calculations for one-stage experiments only. The sample spaces for one-stage experiments are usually quite simple and much easier to visualize than those for multiple-stage experiments. Hereafter, in calculating probabilities, the techniques based on formula (6) and single-stage experiments will be used almost exclusively in order to avoid the time consuming method based on applying formula (2) directly to the sample space of the entire experiment. It should be understood, however, that it is always possible to calculate any type of probability that may arise by working exclusively with the original sample space. A student should occasionally work a problem in this manner to test his understanding of basic concepts. It would be well for him, for example, to solve the preceding illustrative problem by constructing the sample space for that two-stage experiment.

As another illustration of the multiplication rule, calculate the probability of getting 2 prizes in taking 2 punches on a punch board which contains 5 prizes and 20 blanks. If A_1 denotes the event of getting a prize on the first punch and A_2 the event of getting a prize on the second punch, then formula (6) gives

$$P\{A_1 \text{ and } A_2\} = \frac{5}{25} \cdot \frac{4}{24} = \frac{1}{30}.$$

The value of $P\{A_2|A_1\} = \frac{4}{24}$ arises from the fact that since the first punch yielded a prize there are only 4 prizes left and only 24 punches left.

As a final illustration, consider the problem of calculating the probability of getting two aces in drawing two cards from a bridge deck. If A_1 denotes the event of getting an ace on the first drawing and A_2 the event of getting an ace on the second drawing, then formula (6) gives

$$P\{A_1 \text{ and } A_2\} = \frac{4}{52} \cdot \frac{3}{51} = \frac{1}{221}.$$

The value of $P\{A_2|A_1\} = \frac{3}{51}$ because there are only three aces left after the first drawing and only 51 cards remaining.

If the events A_1 and A_2 are such that the probability that A_2 will occur does not depend upon whether or not A_1 occurs, then A_2 is said to be independent of A_1 and one can write

$$P\{A_2|A_1\} = P\{A_2\}.$$

For this case, the multiplication rule reduces to

$$P\{A_1 \text{ and } A_2\} = P\{A_1\}P\{A_2\}.$$

Since the event (A_1 and A_2) is the same as the event (A_2 and A_1), A_1 and A_2 may be interchanged in (6) to give

$$P\{A_1 \text{ and } A_2\} = P\{A_2\}P\{A_1|A_2\}.$$

Comparing the right sides of these two formulas shows that $P\{A_1|A_2\} = P\{A_1\}$. This demonstrates that A_1 is independent of A_2 when A_2 is independent of A_1. Because of this mutual independence, it is proper to say that A_1 and A_2 are independent, without specifying which is independent of the other. As a result,

When A_1 and A_2 are independent,

(7) $$P\{A_1 \text{ and } A_2\} = P\{A_1\}P\{A_2\}.$$

In view of this result, one can state that two events are independent if, and only if, the probability of their joint occurrence is equal to the product of their individual probabilities. This formula generalizes in the obvious manner to more than two independent events. Although it is easy to state the condition that must be satisfied if two events are to be independent, it is not always so easy in real life to decide whether two events are independent. As a rather far-fetched example, suppose the two events are A_1: the stock market will rise next week, and A_2: a stockholder will catch a cold next week. It would seem obvious that the probability that one of these events will occur would be the same whether or not the other event occurred, hence that these are independent events. However, if it should happen that stocks rose considerably in value, the chances are that many stockholders might go out to celebrate their good fortune and thereby increase their chances of catching a cold during that week, in which case these events would not be independent in a probability sense.

In games of chance, such as roulette, it is always assumed that consecutive plays are independent events. If one were not willing to accept this assumption, then one would be forced to assume that the roulette wheel possessed a memory or that the operator of the wheel was secretly manipulating it.

As illustrations of the application of the preceding rules of probability, consider a few simple card problems.

Two cards are drawn from an ordinary deck of 52 cards, the first card drawn being replaced before the second card is drawn.

1.   What is the probability that both cards will be spades? Let A_1 denote the event of getting a spade on the first draw, and A_2 the event of getting a spade on the second draw. Since the first card drawn is replaced, the probability of getting a spade on the second draw should not depend upon whether or not a

spade was obtained on the first draw; hence A_2 may be assumed to be independent of A_1. Formula (7) will then give

$$P\{A_1 \text{ and } A_2\} = \frac{13}{52} \cdot \frac{13}{52} = \frac{1}{16}.$$

2. What is the probability that the cards will be either two spades or two hearts? Let B_1 be the event of getting two spades, and B_2 the event of getting two hearts. Then, from the preceding result, it follows that

$$P\{B_1\} = P\{B_2\} = \frac{1}{16}.$$

Since the events B_1 and B_2 are mutually exclusive and the problem is to calculate the probability that either B_1 or B_2 will occur, formula (3) applies; hence

$$P\{B_1 \text{ or } B_2\} = \frac{1}{16} + \frac{1}{16} = \frac{1}{8}.$$

As before, let two cards be drawn from a deck, but this time the first card drawn will not be replaced.

3. What is the probability that both cards will be spades? Here A_2 is not independent of A_1 because if a spade is obtained on the first draw the chances of getting a spade on the second draw will be smaller than if a nonspade had been obtained on the first draw. For this problem formula (6) must be used. Then

$$P\{A_1 \text{ and } A_2\} = \frac{13}{52} \cdot \frac{12}{51} = \frac{1}{17}.$$

The second factor is $\frac{12}{51}$ because there are only 51 cards after the first drawing, all of which are assumed to possess the same chance of being drawn, and there are only 12 spades left.

4. As a final illustration that does not involve games of chance and that involves more than two independent events, consider the following problem. Assuming that the ratio of male children is $\frac{1}{2}$ (which is only approximately true), find the probability that in a family of 6 children (a) all the children will be of the same sex, (b) 5 of the children will be boys and 1 will be a girl.

(a) Let A_1 be the event that all the children will be boys and A_2 the event that they will all be girls. Because A_1 and A_2 are mutually exclusive events,

$$P\{A_1 \text{ or } A_2\} = P\{A_1\} + P\{A_2\}.$$

Since the six individual births may be assumed to be six independent events with respect to the sex of the child, it follows by using the more general version of the multiplication formula (7) that

$$P\{A_1\} = P\{A_2\} = \left(\frac{1}{2}\right)^6.$$

Hence,

$$P\{A_1 \text{ or } A_2\} = \left(\frac{1}{2}\right)^6 + \left(\frac{1}{2}\right)^6 = \frac{1}{32}.$$

(b) Let A_1 be the event that the oldest child is a girl and the others are boys, A_2 the event that the second oldest is a girl and the others are boys, and similarly for events A_3, A_4, A_5, A_6. Since the event of having 5 boys and 1 girl will occur if, and only if, one of the six mutually exclusive events $A_1, \ldots, A_6$ occurs, it follows from (3) that

$$P\{5 \text{ boys and 1 girl}\} = P\{A_1\} + \cdots + P\{A_6\}.$$

But

$$P\{A_1\} = \cdots = P\{A_6\} = \left(\frac{1}{2}\right)^6.$$

Hence

$$P\{5 \text{ boys and 1 girl}\} = 6\left(\frac{1}{2}\right)^6 = \frac{3}{32}.$$

Although the preceding rules of probability were derived on the assumption that all the possible outcomes of the experiment in question were expected to occur with the same relative frequency, the rules hold for more general experiments. They can even be applied to events related to experiments involving an infinite number of possible outcomes. These more general experiments are considered in Chapters 4 and 5.

5.   BAYES' FORMULA

There is a certain class of important problems based on the application of formula (5) that lead to rather involved computations; therefore it is convenient to have a formula for solving such problems in a systematic manner. These problems may be illustrated by the following academic one. Suppose a box contains 2 red balls and 1 white ball and a second box contains 2 red balls and 2 white balls. One of the boxes is selected by chance and a ball drawn from it. If the drawn ball is red, what is the probability that it came from the first box? Let A_1 denote the event of choosing the first box and let A_2 denote the event of drawing a red ball. Then the problem is to calculate the conditional probability $P\{A_1|A_2\}$. This will be done by the use of formula (5) with A_1 and A_2 interchanged in that formula. Since the phrase by chance is understood to mean that each box has the same probability of being chosen, it follows that the probability of drawing the first box is $\frac{1}{2}$, and that of drawing the second box is the same. The calculation of the numerator term in (5) can be accomplished by using formula (6) in the order of events listed there. Thus

$$P\{A_2 \text{ and } A_1\} = P\{A_1 \text{ and } A_2\} = \frac{1}{2}\cdot\frac{2}{3} = \frac{1}{3}.$$

The denominator, $P\{A_2\}$, can be calculated by considering the two mutually exclusive ways in which A_2 can occur, namely, getting the first box and then a red ball or getting the second box and then a red ball. By formula (3), $P\{A_2\}$ will be given by the sum of the probabilities of those two mutually exclusive possibilities; hence

$$P\{A_2\} = \frac{1}{2}\cdot\frac{2}{3} + \frac{1}{2}\cdot\frac{2}{4} = \frac{7}{12}.$$

Application of the modified version of formula (5) then yields the desired result, namely,

$$P\{A_1|A_2\} = \frac{1}{3}\bigg/\frac{7}{12} = \frac{4}{7}.$$

This problem could have been worked very easily by looking at the sample space for the experiment; however, the objective here is to work with formula (5) and attempt to obtain a formula for treating more complicated problems of the type of the present one.

The foregoing problem is a special case of problems of the following type. One is given a two-stage experiment. The first stage can be described by stating that exactly one of, say, k possible outcomes must occur when the complete experiment is performed. Those possible outcomes will be denoted by $e_1, e_2, \ldots, e_k$. In the second stage there are, say, m possible outcomes, exactly one of which must occur. These will be denoted by $o_1, o_2, \ldots, o_m$. The values of the probabilities for each of the possible outcomes $e_1, e_2, \ldots, e_k$ are given. As before, they will be denoted by $P\{e_1\}, P\{e_2\}, \ldots, P\{e_k\}$. The values of all the conditional probabilities of the type $P\{o_j|e_i\}$, which represents the probability that the second stage event o_j will occur when it is known that the first-stage event e_i occurred, are also given. The problem is to calculate the probability that the first-stage event e_i occurred when it is known that the second-stage event o_j occurred. This conditional probability would be written $P\{e_i|o_j\}$. For simplicity of notation, the calculations will be carried out for $P\{e_1|o_1\}$; the calculations for any other pair would be the same.

In terms of the present notation, formula (5) assumes the form

(8)
$$P\{e_1|o_1\} = \frac{P\{e_1 \text{ and } o_1\}}{P\{o_1\}}.$$

Formula (6) gives

(9)
$$P\{e_1 \text{ and } o_1\} = P\{e_1\}P\{o_1|e_1\}.$$

From the information given in this problem it will be observed that the two probabilities on the right side of (9) are known; therefore the numerator in (8) can be obtained from (9). The value of $P\{o_1\}$ in (8) can be computed by considering all the mutually exclusive ways in which o_1 can occur in conjunction with the first stage of the experiment. The second-stage event o_1 will occur if the first-stage event e_1 occurs and then o_1 occurs, or if the first-stage event e_2 occurs and then o_1 occurs, ..., or if the first-stage event e_k occurs and then o_1 occurs. If e_1 is replaced by e with the appropriate subscript in (9), that formula can be used to calculate the probability for each of these mutually exclusive possibilities. Application of formula (3) to these k mutually exclusive events then yields the formula

$$P\{o_1\} = P\{e_1\}P\{o_1|e_1\} + P\{e_2\}P\{o_1|e_2\} + \cdots + P\{e_k\}P\{o_1|e_k\}.$$

This can be written in condensed form by using the summation symbol Σ as follows:

$$P\{o_1\} = \sum_{i=1}^{k} P\{e_i\}P\{o_1|e_i\}.$$

Here e_i denotes event number i in the set $e_1, e_2, \ldots, e_k$ and the summation symbol tells one to add all the probabilities of the type displayed beginning with e_1 and ending with e_k. This result together with (9) when applied to (8) will give the desired formula which is known as Bayes' formula.

(10)   **Bayes' Formula.** $P\{e_1|o_1\} = \dfrac{P\{e_1\}P\{o_1|e_1\}}{\displaystyle\sum_{i=1}^{k} P\{e_i\}P\{o_1|e_i\}}.$

Returning to the problem that was solved earlier without this formula, one will observe that there were two events e_1 and e_2 in the first stage corresponding to choosing the first or the second box and that $P\{e_1\} = P\{e_2\} = \frac{1}{2}$. The second stage also consisted of two events o_1 and o_2 corresponding to obtaining a red or a white ball. The conditional probabilities of obtaining a red ball based on what transpired at the first stage were given by $P\{o_1|e_1\} = \frac{2}{3}$ and $P\{o_1|e_2\} = \frac{2}{4}$. It will be observed that the substitution of these values in (10) yields the result that was obtained before.

Consider now a more practical application of this formula. Suppose a test for detecting a certain rare disease has been perfected that is capable of discovering the disease in 97 per cent of all afflicted individuals. Suppose further that when it is tried on healthy individuals, 5 per cent of them are incorrectly diagnosed as having the disease. Finally, suppose that when it is tried on individuals who have certain other milder diseases, 10 per cent of them are incorrectly diagnosed. It is known that the percentages of individuals of the three types being considered here in the population at large are 1 per cent, 96 per cent, and 3 per cent, respectively. The problem is to calculate the probability that an individual, selected at random from the population at large and tested for the rare disease, actually has the disease if the test indicates he is so afflicted.

Here there are three events e_1, e_2, and e_3 in the first stage corresponding to the three types of individuals in the population. Their corresponding probabilities are $P\{e_1\} = .01$, $P\{e_2\} = .96$, and $P\{e_3\} = .03$. There are two events o_1 and o_2 in the second stage corresponding to whether the test claims that the individual has the disease or not. The conditional probabilities are given by $P\{o_1|e_1\} = .97$, $P\{o_1|e_2\} = .05$, and $P\{o_1|e_3\} = .10$. In terms of the present notation, the problem is to calculate $P\{e_1|o_1\}$. A direct application of formula (10) based on the preceding probabilities will supply the answer, namely,

$$P\{e_1|o_1\} = \frac{(.01)(.97)}{(.01)(.97) + (.96)(.05) + (.03)(.10)} = .16.$$

This result may seem rather surprising because it shows that only 16 per cent of the individuals whom the test would indicate have the disease actually do have it when the test is applied to the population at large. The 84 per cent who were falsely diagnosed might resent the temporary mental anguish caused by their belief that they had the disease before further tests revealed the falsity of the diagnosis. They might also resent the necessity of having been required to undergo further tests when it turned out that those tests were really unnecessary. A calculation such as the preceding one might therefore cause authorities to ponder a bit before advocating mass testing.

▶6.   COUNTING TECHNIQUES—TREES

It is convenient in solving some of the more difficult problems involving two or more stage experiments to have systematic methods for calculating the compound event probabilities that arise. A pictorial method that has proved particularly useful is based on what is known as a *probability tree*.

For the purpose of illustrating how such a tree is constructed, consider once more the first example to which formula (6) was applied. A box contains three red, two black, and one green ball and two balls are to be drawn. This is a two-stage experiment for which the various possibilities that can occur may be represented by a horizontal tree such as that shown in Fig. 10. Each stage of a multiple-stage experiment has as many branches as there are possibilities at that stage. Here there are three main branches at the first stage and three branches at each of the second stages, except for the last one where there are only two branches because it is impossible to obtain a green ball at the second drawing if a green ball is obtained on the first drawing. The total number of terminating branches in such a tree gives the total number of possible outcomes in the

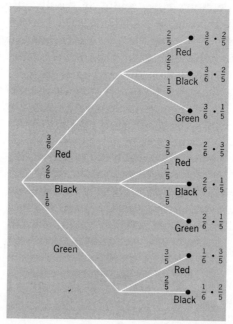

FIGURE 10.   A probability tree.

compound experiment, and therefore the end points of those branches may be treated as the sample points of a sample space.

The probability that is attached to any branch of the tree is the conditional probability that the event listed under that branch will occur, subject to the condition that the preceding branch events all occurred. Thus, the $\frac{2}{5}$ listed above the top terminal branch is the conditional probability that a red ball will be obtained on the second drawing if a red ball was obtained on the first drawing. The probability listed at the end of a terminal branch is the probability of obtaining the sequence of events that are required to arrive at that terminal point and is obtained by multiplying the probabilities associated with the branches leading to that terminal point. Thus, the first terminal probability $\frac{3}{6} \cdot \frac{2}{5}$ is the probability of obtaining a red ball at the first drawing times the conditional probability of doing so at the second drawing. By means of this tree and its probabilities it is relatively easy to answer various probability questions.

Probability trees yield a simple pictorial method for calculating probabilities for which Bayes' formula would normally be employed. As an example, consider the earlier problem that was used to motivate Bayes' formula, namely the problem of calculating $P(A_1 | A_2)$ where A_1 is the event of selecting Box 1 and A_2 is the event of getting a red ball. The tree corresponding to this two-stage experiment is shown in Fig. 11 with the proper probabilities attached to each branch and to each of the four sample points.

Now the topmost branch corresponds to the compound event A_1 and A_2; therefore $P(A_1 \text{ and } A_2) = \frac{1}{2} \cdot \frac{2}{3}$. Furthermore, it follows from definition (1) that

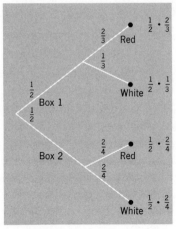

FIGURE 11. A tree for
a Bayesian problem.

$P(A_2) = \frac{1}{2} \cdot \frac{2}{3} + \frac{1}{2} \cdot \frac{2}{4}$, because the event A_2 consists of the two sample points associated with the word "Red." Hence, by formula (5)

$$P\{A_1 | A_2\} = \frac{P\{A_1 \text{ and } A_2\}}{P\{A_2\}} = \frac{\frac{1}{2} \cdot \frac{2}{3}}{\frac{1}{2} \cdot \frac{2}{3} + \frac{1}{2} \cdot \frac{2}{4}} = \frac{4}{7}.$$

The technique is now seen to be the following one. After constructing the probability tree, select the terminal branch that corresponds to the occurrence of both A_1 and A_2. Then divide the probability associated with that terminal branch by the sum of the probabilities of all the terminal branches that are associated with the event A_2.

This technique will also be applied to the second illustration associated with Bayes' formula. First, it is necessary to construct a tree such as that shown in Fig. 12. To obtain $P\{e_1 | o_1\}$ it now suffices to divide the probability $(.01)(.97)$ associated with the top terminal branch by the sum of the probabilities of the terminal branches associated with the letter o_1. Thus

$$P\{e_1 | o_1\} = \frac{(.01)(.97)}{(.01)(.97) + (.96)(.05) + (.03)(.10)} = .16.$$

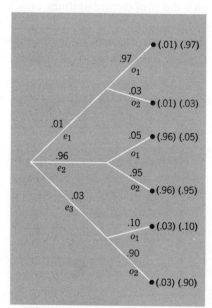

FIGURE 12. A tree for a Bayesian problem.

▶7.   COUNTING TECHNIQUES—COMBINATIONS

If there are many stages to an experiment and several possibilities at each stage, the probability tree associated with the experiment would become too large to be manageable. For such problems the counting of sample points is simplified by means of algebraic formulas. Toward this objective, consider a two-stage experiment for which there are r possibilities at the first stage and s possibilities at the second stage. A tree to represent this experiment is shown in Fig. 13. Since each of the r main branches has s terminal branches attached to it, the total number of possibilities here is rs, and therefore this would be the number of sample points in the sample space for this two-stage experiment. If a third stage

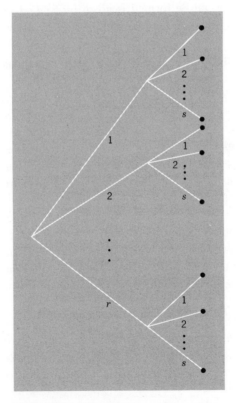

FIGURE 13.   A tree for a two-stage experiment.

with t possibilities were added, the total number of sample points would become rst. This can be extended in an obvious manner to any number of stages.

Now consider the application of this counting method to the problem of determining in how many ways it is possible to select r objects from n distinct objects. Toward this objective, first consider the particular problem of determining how many three-letter words can be formed from the five letters a, b, c, d, e if a letter may be used only once in a given word and if any set of three letters is called a three-letter word.

Forming a three-letter word may be thought of as a three-stage experiment in which the first stage is that of choosing the first letter of the word. In this problem there are five possibilities for the first stage, but only four possibilities for the second stage because the letter chosen at the first stage is not available at the second stage. There are only three possibilities left at the third stage; therefore, according to the preceding counting method, the total number of three-letter words that can be formed is given by $5 \cdot 4 \cdot 3 = 60$.

The preceding problem will be modified slightly by asking: How many committees consisting of three individuals can be selected from a group of five individuals? If the letters a, b, c, d, e are associated with the five individuals, then a three-letter word will correspond to a committee of three; however, two words using the same three letters but in a different order will correspond to the same committee. For example, *ace* and *cea* are distinct words but do not represent different committees. Since the three letters a, c, and e will produce only one committee but will give rise to $3 \cdot 2 \cdot 1 = 6$ distinct three-letter words, and this will be true for every selection of three letters, it follows that there will be only $\frac{1}{6}$ as many committees of three as there are three-letter words; hence there will be $\frac{60}{6} = 10$ such committees.

Now suppose one is given n distinct objects, say n distinct letters a, b, c, ..., and that r of those objects are to be chosen to form r-letter words. This can be treated as an r-stage experiment in which there are n possibilities at the first stage, $n - 1$ possibilities at the second stage, etc. Then, the total number of words that can be formed is given by the formula

(11) $$_nP_r = n(n - 1)(n - 2) \cdots (n - r + 1).$$

The symbol $_nP_r$ is called the number of permutations of n objects taken r at a time. An arrangement along a line of a set of objects is called a *permutation* of those objects; therefore, an r-letter word is a permutation of the r letters used to construct the word.

The corresponding problem of counting the number of committees consisting of r individuals that can be selected from n individuals is readily solved by means of the preceding formula. Just as in the preceding special problem, it suffices to realize that a committee is concerned only with which r letters are selected and not in how they are arranged along a line. Since r distinct letters can be arranged along a line to produce $r(r - 1)(r - 2) \cdots 1$ distinct words but only one committee, the number of words must be divided by $r(r - 1)(r - 2) \cdots 1$ to give the number of distinct committees. Thus, the total number of committees that can be formed is given by the formula

(12) $$\binom{n}{r} = \frac{n(n - 1)(n - 2) \cdots (n - r + 1)}{r(r - 1)(r - 2) \cdots 1}$$

The symbol $\binom{n}{r}$ is called the number of combinations of n things taken r at a time. A *combination* of r objects is merely a selection of r objects without regard to the order in which they are selected or arranged after selection. A permutation of r objects may be thought of as the outcome of a two-stage experiment in which the first stage consists in choosing a combination of r objects and then arranging that combination along a line.

A convenient symbol to use in connection with formula (12) is the *factorial* symbol, which consists of an exclamation mark after an integer, indicating that the number concerned should be multiplied by all the positive integers smaller than it. Thus $4! = 4 \cdot 3 \cdot 2 \cdot 1$ and $r! = r(r - 1)(r - 2) \cdots 1$. In order to allow r to assume the value 0 in the following formulas, one defines $0! = 1$. Formula (12) can therefore be written in the form

$$\binom{n}{r} = \frac{n(n - 1) \cdots (n - r + 1)}{r!}.$$

The numerator of this expression can also be expressed in terms of factorial notation by observing that

$$\frac{n!}{(n - r)!} = \frac{n(n - 1) \cdots (n - r + 1)(n - r)(n - r - 1) \cdots 1}{(n - r)(n - r - 1) \cdots 1}$$
$$= n(n - 1) \cdots (n - r + 1).$$

By using factorial symbols in this manner, formula (12) can be written in the compact form

(13) $$\binom{n}{r} = \frac{n!}{r!(n - r)!}.$$

The usefulness of formula (12), or (13), for counting purposes will be illustrated in the calculation of the probability of getting five spades in a hand of five cards drawn from a deck of 52 playing cards. Here the total number of possible outcomes corresponds to the total number of five-card hands that can be formed from a deck of 52 distinct cards. Since arrangement is of no interest here, this is a combination counting problem. The total number of possible hands is, according to formula (12), given by

$$\binom{52}{5} = \frac{52 \cdot 51 \cdot 50 \cdot 49 \cdot 48}{5 \cdot 4 \cdot 3 \cdot 2 \cdot 1}.$$

The number of outcomes that correspond to the occurrence of the desired event is equal to the number of ways of selecting five spades from thirteen spades. This is given by

$$\binom{13}{5} = \frac{13 \cdot 12 \cdot 11 \cdot 10 \cdot 9}{5 \cdot 4 \cdot 3 \cdot 2 \cdot 1}.$$

The desired probability is given by the ratio of these two numbers; hence it is equal to

$$\frac{13 \cdot 12 \cdot 11 \cdot 10 \cdot 9}{52 \cdot 51 \cdot 50 \cdot 49 \cdot 48} = \frac{33}{66640} = .0005.$$

The moral seems to be that one should not expect to obtain a five-card poker hand containing only spades. Even if one settles for a hand containing five cards of the same suit, the probability is only four times as large, namely .002, which is still hopelessly small.

As a final illustration of the use of these formulas, consider the following problem. If a list of 20 individuals who volunteered to supply blood, when it is needed for a transfusion, has 15 individuals of type B blood and if 3 individuals are selected at random from the list, what is the probability that (a) all three will be of type B, (b) 2 will be of type B and 1 will not, (c) at least 1 will be of type B?

(a) The total number of possible outcomes is the total number of ways of choosing 3 individuals from a group of 20 individuals, which is given by $\binom{20}{3}$. The number of outcomes that correspond to the occurrence of the desired event is the number of ways of choosing 3 individuals from a group of 15 individuals, which is given by $\binom{15}{3}$. The probability of this favorable event occurring is therefore given by the ratio

$$\frac{\binom{15}{3}}{\binom{20}{3}} = \frac{15 \cdot 14 \cdot 13}{20 \cdot 19 \cdot 18} = \frac{91}{228} = .40.$$

(b) Here the number of favorable outcomes is given by the number of ways of choosing 2 individuals from a group of 15, multiplied by the number of ways of choosing 1 individual from a group of 5, which gives $\binom{15}{2}\binom{5}{1}$. The probability of this event occurring is therefore given by the ratio

$$\frac{\binom{15}{2}\binom{5}{1}}{\binom{20}{3}} = \frac{15\cdot 14\cdot 5\cdot 3}{20\cdot 19\cdot 18} = \frac{35}{76} = .46.$$

(c) This event will occur if the event of getting 0 such individuals does not occur. The latter probability is given by $\binom{5}{3}\Big/\binom{20}{3}$; hence the desired probability is given by

$$1 - \frac{\binom{5}{3}}{\binom{20}{3}} = 1 - \frac{5\cdot 4\cdot 3}{20\cdot 19\cdot 18} = 1 - \frac{1}{114} = \frac{113}{114} = .99.$$

8.   ADDITIONAL ILLUSTRATIONS

Just as for the preceding chapter, and also the following ones, this last section is designed as a review of the problem solving techniques presented in the chapter. If it is to serve as a review, the student should attempt to solve these problems before inspecting the solutions. If it is to serve as additional illustrations as one proceeds from section to section, then the section number associated with each part of a problem will indicate to which section that problem belongs.

1.   Each of two dice has been altered by having its one-spot changed to a two-spot. As a result, each die will contain two two's but no one's. The two dice are then rolled once. Assuming that the two dice can be distinguished, solve the following problems: (a) 1 Construct a sample space for the experiment. (b) 1 Assign probabilities to the points of the sample space. (c) 2 Using definition (1) calculate the probability i) of getting a total of 6 points on the two dice, ii) that at least one of the dice will show a 2, iii) that at least one of the dice will show a 2 if it is known that no number larger than 4 was obtained on either die. The solutions follow.

(a) The sample space is conveniently represented by the following 25 points which have been labeled by indicating the outcome on each die.

22	32	42	52	62
23	33	43	53	63
24	34	44	54	64
25	35	45	55	65
26	36	46	56	66

(b) Since the two dice are assumed to behave in the same manner as two normal dice for which the sample space is given in Table 1 and for which each point was assigned the probability $\frac{1}{36}$, the probabilities to be assigned here should be in agreement with those probabilities. Hence, since the event 22 here corresponds to the composite event consisting of the simple events 11, 12, 21, and 22 for Table 1, the probability $\frac{4}{36}$ should be assigned to the point labeled 22 of the present sample space. Each of the remaining points of this space that has a 2 in its label should be assigned the probability $\frac{2}{36}$ because there are two simple events of Table 1 that produced it. All other points should be assigned the probability $\frac{1}{36}$ because they do not differ from the corresponding ones in Table 1.

(c) i) If A denotes this event, it is seen by inspecting the sample space in (a) that the three points 24, 33, and 42 constitute the composite event A; hence applying definition (1) and using the probabilities assigned to those points in (b), it follows that

$$P\{A\} = \frac{2}{36} + \frac{1}{36} + \frac{2}{36} = \frac{5}{36}.$$

ii) If B denotes this event, it is seen that B consists of the simple events 22, 23, 24, 25, 26, 32, 42, 52, and 62. From (b), the point 22 was assigned the probability $\frac{4}{36}$ and the remaining points the probability $\frac{2}{36}$; consequently, since there are eight of the latter,

$$P\{B\} = \frac{4}{36} + 8\left(\frac{2}{36}\right) = \frac{5}{9}.$$

iii) If C denotes the event of getting at least one 2, and D denotes the event that no number larger than a 4 shows, it is seen that D consists of the simple events 22, 23, 24, 32, 33, 34, 42, 43, and 44. Of these points, only 22, 23, 24, 32, and 42 also lie in C; hence it follows from the assignment of probabilities in (b) that

$$P\{D\} = \frac{4}{36} + 4\left(\frac{2}{36}\right) + 4\left(\frac{1}{36}\right) = \frac{4}{9}$$

$$P\{C \text{ and } D\} = \frac{4}{36} + 4\left(\frac{2}{36}\right) = \frac{1}{3}$$

$$P\{C|D\} = \frac{1}{3} \Big/ \frac{4}{9} = \frac{3}{4}.$$

2. A purse contains two pennies, three nickels, and four dimes. Two coins are to be selected at random from this purse. Using the addition and multiplication formulas, calculate the probability that (a) 4 both coins will be pennies, (b) 4 one coin will be a penny and the other will be a dime, (c) 4 both coins will be the same kind, (d) 4 neither coin will be a nickel, (e) 4 at least one nickel will be obtained. The solutions follow.

(a) Applying formula (6) and considering the experiment in two stages gives

$$P\{pp\} = \frac{2}{9} \cdot \frac{1}{8} = \frac{1}{36}.$$

(b) This event will occur if either of the two mutually exclusive events pd or dp occurs. Application of formulas (6) and (3) then gives the result

$$P\{pd\} + P\{dp\} = \frac{2}{9} \cdot \frac{4}{8} + \frac{4}{9} \cdot \frac{2}{8} = \frac{2}{9}.$$

(c) This event will occur if one of the three mutually exclusive events pp, nn, or dd occurs; hence the answer is given by

$$P\{pp\} + P\{nn\} + P\{dd\} = \frac{2}{9} \cdot \frac{1}{8} + \frac{3}{9} \cdot \frac{2}{8} + \frac{4}{9} \cdot \frac{3}{8} = \frac{5}{18}.$$

(d) There are three nickels and six non-nickel coins; hence the probability of getting two non-nickel coins is $\frac{6}{9} \cdot \frac{5}{8} = \frac{5}{12}$.

(e) The probability of obtaining at least one nickel is equal to 1 minus the probability of getting two non-nickel coins; hence from part (d) the answer must be $1 - \frac{5}{12} = \frac{7}{12}$.

3. A box contains the following five cards: the ace of spades, the ace of clubs, the two of hearts, the two of diamonds, and the three of spades. An ace is considered as a one. Spades and clubs are black cards, whereas hearts and diamonds are red cards. Two cards are to be drawn from this box without replacing

the first card drawn before the second drawing. Using the addition and multiplication formulas, calculate the probability that (a) 4 both cards will be red, (b) 4 the first card will be an ace and the second card will be a two, (c) 4 both cards will be the same color, (d) 4 one card will be a spade and the other will be a club, (e) 4 a total of 4 on the two cards will be obtained, (f) 5 exactly one ace will be obtained if it is known that both cards are black, (g) 5 the first card drawn was a spade if the second card drawn turned out to be the ace of spades. Work parts (f) and (g) by employing formula (10). The solutions follow.

(a) Applying formula (6) and considering the experiment in two stages,

$$P\{RR\} = \frac{2}{5} \cdot \frac{1}{4} = \frac{1}{10}.$$

(b) Applying formula (6),

$$P\{A_2\} = \frac{2}{5} \cdot \frac{2}{4} = \frac{1}{5}.$$

(c) The two events RR and BB constitute the two mutually exclusive ways in which the desired event can occur; hence applying formula (3), the desired probability is given by

$$P\{RR \text{ or } BB\} = P\{RR\} + P\{BB\} = \frac{2}{5} \cdot \frac{1}{4} + \frac{3}{5} \cdot \frac{2}{4} = \frac{2}{5}$$

(d) The two events SC and CS will satisfy; hence

$$P\{SC \text{ or } CS\} = P\{SC\} + P\{CS\} = \frac{2}{5} \cdot \frac{1}{4} + \frac{1}{5} \cdot \frac{2}{4} = \frac{1}{5}.$$

(e) A total of 4 will be obtained if both cards are two's, or if one card is a three and the other is a one. If a subscript is used to denote the number on a card, the events that will satisfy are the following ones: H_2D_2, D_2H_2, S_3S_1, S_1S_3, S_3C_1, C_1S_3. Since these constitute the mutually exclusive ways in which the desired event can occur and since each of these possesses the same probability, namely $\frac{1}{5} \cdot \frac{1}{4} = \frac{1}{20}$, it follows that

$$P\{4 \text{ total}\} = \frac{6}{20} = \frac{3}{10}.$$

(f) Let A_1 denote the event that both cards will be black and A_2 the event that exactly one ace will be obtained. Then $P\{A_2|A_1\}$ is the probability needed to solve the problem. From formula (5), this requires the computation of $P\{A_1\}$ and $P\{A_1 \text{ and } A_2\}$. First $P\{A_1\} = \frac{3}{5} \cdot \frac{2}{4} = \frac{3}{10}$. Next, both A_1 and A_2 will occur if one of the following mutually exclusive events occurs: S_1S_3, S_3S_1, C_1S_3, S_3C_1. Since each of these events has the probability $\frac{1}{5} \cdot \frac{1}{4} = \frac{1}{20}$ and there are four of them, it follows that $P\{A_1 \text{ and } A_2\} = \frac{4}{20} = \frac{1}{5}$. Hence,

$$P\{A_2|A_1\} = \frac{1}{5}\bigg/\frac{3}{10} = \frac{2}{3}.$$

(g) Let A_1 denote the event of getting a spade on the first draw and let A_2 denote the event of getting the ace of spades on the second draw. The problem is to calculate $P\{A_1|A_2\}$. Here one can consider five events, e_1, e_2, e_3, e_4, and e_5, to represent the possible outcomes of the first stage of the experiment where these symbols denote the events of getting S_1, C_1, H_2, D_2, and S_3, respectively. Since all these events have the same probability of occurring $P\{e_1\} = \cdots = P\{e_5\} = \frac{1}{5}$. Next, one can consider two events, o_1 and o_2, to represent the possible outcomes of the second stage of the experiment, where o_1 denotes the event of getting the ace of spades and o_2 the event of not getting the ace of spades on the second draw. Here

$$P\{o_1|e_1\} = 0, P\{o_1|e_2\} = P\{o_1|e_3\} = P\{o_1|e_4\} = P\{o_1|e_5\} = \frac{1}{4}.$$

Now the event A_1 will occur if either of the two mutually exclusive events e_1 or e_5 occurs; hence

$$P\{A_1|A_2\} = P\{e_1|A_2\} + P\{e_5|A_2\}.$$

In the present notation $A_2 = o_1$; therefore

$$P\{A_1|A_2\} = P\{e_1|o_1\} + P\{e_5|o_1\}.$$

Each of these probabilities will be calculated by means of formula (10). Since they have the same denominator, the result can be written in the form

$$P\{A_1|A_2\} = \frac{\frac{1}{5}\cdot 0 + \frac{1}{5}\cdot\frac{1}{4}}{\frac{1}{5}(0 + \frac{1}{4} + \frac{1}{4} + \frac{1}{4} + \frac{1}{4})} = \frac{1}{4}.$$

This problem and its solution calls for a few comments. No attempt was made in the solution to use what might be called common sense because the problem was treated as an exercise in the application of formula (10). Thus, the fact that e_1 could not have occurred if o_1 occurs would eliminate the necessity of considering the probability $P\{e_1|o_1\}$. Furthermore, if one looks at the implication of the restriction that o_1 must occur, which means that the second card must be the ace of spades, it would suggest that this is equivalent to considering experiments in which the ace of spades is removed and saved for the second drawing, in which case the probability of getting a black card on the first draw is $\frac{1}{4}$. Thus, one can often arrive at correct answers very quickly by visualizing simpler experiments which are expected to yield equivalent results. This is not the same as solving the original problem by means of formulas, however, which was required here.

EXERCISES

Section 1

1. List all the possible outcomes if a coin is tossed 4 times.

2. A box contains one red, one black, and one green ball. Two balls are to be drawn from this box without replacing the first ball drawn before the second drawing. Construct a sample space for this experiment similar to Table 1.

3. What probabilities should be assigned to the points of the sample space corresponding to the experiment of problem 1?

4. What probabilities would you assign to the points of the sample space of problem 2? What assignment would you have made if the first ball was returned to the box before the second drawing?

5. A box contains 2 black and 1 white ball. Two balls are to be drawn from this box. Construct a sample space for this experiment (a) using 6 points, (b) using 3 points.

6. What probabilities would you assign to the points of the two sample spaces constructed in problem 5?

7. What expected relative frequencies would you guess should be assigned to the possible outcomes of an experiment consisting of selecting a female student at random and noting whether she would be rated as a blonde, redhead, or brunette? How would you go about improving on your guess?

8. If you were interested in studying the fluctuations of the stock market over a period of time, would you expect the relative frequency of rises to be about the same as the relative frequency of declines? Would it be possible for this to be true and yet have the stock market rise in value regularly over a period of a year?

9. Let e_1, e_2, and e_3 denote the events of getting a digit less than 4, getting a digit between 4 and 6 inclusive, and getting a digit larger than 6, respectively, when selecting a digit from the table of random digits. (a) Construct a sample space for this experiment and assign probabilities to the points. (b) Perform the experiment 200 times and calculate the experimental relative frequencies for the three events to see whether your model seems to be a realistic one.

Section 2

10. A box contains 4 red, 3 black, 2 green, and 1 white ball. A ball is drawn from the box and then returned to the box. What is the probability that the ball will be (a) red,

(b) red or black? Now simulate this experiment by means of random numbers by calling the digits 0, 1, 2, 3 red, the digits 4, 5, 6 black, the digits 7, 8 green, and the digit 9 white, and perform the experiment of selecting a digit from the table of random digits 400 times. Let A_1 and A_2 denote the events in parts (a) and (b), respectively, and keep a tabulation of the number of times A_1 and A_2 occurred. Observe whether the mathematical model assumed to hold here seems to be realistic.

11. An honest die is rolled twice. Using Table 1, calculate the probability of getting (a) a total of 5, (b) a total of less than 5, (c) a total that is an even number.

12. An honest coin is tossed 4 times. Using the model of problems 1 and 3, calculate the probability of getting (a) 4 heads, (b) 3 heads and 1 tail, (c) at least 2 heads.

13. For the experiment of rolling two honest dice, calculate the probability that (a) the sum of the numbers will not be 10, (b) neither 1, 2, or 3 will appear, (c) each die will show 2 or more points, (d) the numbers on the two dice will not be the same, (e) exactly one die will show fewer than 3 points.

14. A box contains 2 black, 2 white, and 1 green ball. Two balls are to be drawn from the box. Construct a two-dimensional sample space for this experiment using 20 sample points and assigning equal probabilities to those points. Using this sample space and definition (1) or formula (2), calculate the probability of getting (a) two black balls, (b) one black and one white ball, (c) one black and one green ball.

15. Work problem 14 if the first ball drawn is returned to the box before the second ball is drawn.

Section 3

16. Let A_1 and A_2 be the events associated with Fig. 6. Use that sample space to calculate $P\{A_1 \text{ or } A_2\}$. Since A_1 and A_2 are not mutually exclusive, formula (3) cannot be applied here.

17. With the aid of Fig. 5, show that the addition rule for two events A_1 and A_2 that are not mutually exclusive is given by the formula

$$P\{A_1 \text{ or } A_2\} = P\{A_1\} + P\{A_2\} - P\{A_1 \text{ and } A_2\}.$$

Section 4

18. Two balls are to be drawn from an urn containing 2 white and 4 black balls. (a) What is the probability that the first ball will be white and the second black? (b) What is this probability if the first ball is replaced before the second drawing?

19. Two balls are to be drawn from an urn containing 2 white, 3 black, and 4 green balls. (a) What is the probability that both balls will be green? (b) What is this probability if the first ball is replaced before the second drawing? (c) What is the probability that both balls will be the same color?

20. A box contains 5 coins, 4 of which are honest coins but the fifth of which has heads on both sides. If a coin is selected from the box and then is tossed 2 times, what is the probability that 2 heads will be obtained?

21. The following numbers were obtained from a mortality table based on 100,000 individuals:

Age	Number Alive	Deaths per 1000 During That Year
17	94,818	7.688
18	94,089	7.727
19	93,362	7.765
20	92,637	7.805
21	91,914	7.855

If these numbers are used to define probabilities of death for the corresponding age group and if A, B, and C denote individuals of ages 17, 19, and 21, respectively, calculate the probability that during the year (a) A will die and B will live, (b) A and B will both die, (c) A and B will both live, (d) at least one of A and B will die, (e) at least one of A, B, and C will die.

22. A testing organization wishes to rate a particular brand of table radios. Six radios are selected at random from the stock of radios and the brand is judged to be satisfactory if nothing is found wrong with any of the 6 radios. (a) What is the probability that the brand will be rated as satisfactory if 10 per cent of the radios actually are defective? (b) What is this probability if 20 per cent are defective?

23. Let A_1 and A_2 be the events associated with Fig. 6. Use the general formula of exercise 17 to calculate $P\{A_1 \text{ or } A_2\}$ and compare your answer with that of exercise 16.

Section 5

24. Suppose that 10 per cent of car owners who have an accident will have had at least one other accident during the past year and suppose that a driving simulator test is failed by 80 per cent of such drivers but by only 30 per cent of drivers who had only the one accident. If an individual car owner selected at random from those having had an accident takes the test and fails, what is the probability that he is the type who will have had additional accidents the past year?

25. Assume that there are equal numbers of male and female students in a high school and that the probability is $\frac{1}{5}$ that a male student and $\frac{1}{25}$ that a female student will be a science major. What is the probability that (a) a student selected at random will be a male science student, (b) a student selected at random will be a science student, (c) a science student selected at random will be a male student?

26. Suppose a college aptitude test designed to separate high school students into promising and not-promising groups for college entrance has had the following experience. Among the students who made satisfactory grades in their first year at college 80 per cent passed the aptitude test. Among the students who did unsatisfactory work their first year 30 per cent passed the test. It is assumed that the test was not used for admission to college. If it is known that only 70 per cent of first year college students do satisfactory work, what is the probability that a student who passed the test will be a satisfactory student?

Sections 6 and 7

▶**27.** Four cards are to be drawn from an ordinary deck of 52 cards. (a) What is the probability that all 4 cards will be spades? (b) What is the probability that all 4 cards will be of the same suit? (c) What is the probability that none of the 4 cards will be spades?

▶**28.** If a poker hand of 5 cards is drawn from a deck, what is the probability that it will contain exactly 1 ace?

▶**29.** Suppose the probability is p that the weather (sunshine or cloudy) is the same on any given day as it was on the preceding day. If it is sunny today, what is the probability that it will be sunny the day after tomorrow? Use a tree to assist you.

▶**30.** Employees in a certain firm are given an aptitude test when first employed. Experience has shown that of the 60 per cent who passed the test 80 per cent of them were good workers whereas of the 40 per cent who failed only 30 per cent were rated as good workers. What is the probability that an employee selected at random will be a good worker? Use a tree here.

▶**31.** If a box contains 40 good and 10 defective fuses and 10 fuses are selected, what is the probability that they will all be good? Use combination symbols here.

▶**32.** Find the probability that a poker hand of 5 cards will contain only black cards if it is known to contain at least 4 black cards.

▶**33.** A bridge hand of 13 cards is drawn from a deck of 52 cards. Use combination symbols to calculate the probability that the hand (a) will contain exactly one ace, (b) will contain at least one ace, (c) will contain at least 6 spades, (d) will contain only spades.

Section 8

34. For the experiment of rolling the 2 altered dice of exercise 1 of section 8, calculate the probability that (a) the sum of the numbers will be less than 7, (b) the number 2 will occur, (c) exactly 1 die will show fewer than 3 points.

35. For the third of the two review exercises of section 8 work the following problems by means of the addition and multiplication formulas. Here, however, assume that the first card is returned before the second drawing. What is the probability that (a) both cards will be aces, (b) the ace of spades is certain to be obtained, (c) at least one card will be an ace, (d) at most one card will be an ace, (e) a red ace will not be obtained, (f) the sum of the numbers on the two cards will be less than 4, ▶(g) at least one ace will be obtained if it is known that neither card is a three, ▶(h) the first card was an ace if it is known that the second card is not an ace?

36. Work the review exercises of problem 35 under the assumption that the first card drawn is not returned before the second drawing.

CHAPTER 4

Probability Distributions

Chapter 2 was concerned with sample frequency distributions and their description. This chapter is concerned with population frequency distributions and their properties. A sample frequency distribution is an estimate of the population frequency distribution corresponding to it. If the size of the sample is large, one would expect the sample frequency distribution to be a good approximation of the population frequency distribution. For example, if in a study of dormitory weights one had taken a sample of 400 students and there were only 800 dormitory students on campus, one would have expected the sample and population frequency distributions to be very similar.

In most statistical problems the sample is not large enough to determine the population distribution with much precision. However, there is usually enough information in the sample, together with information obtained from other sources, to suggest the general type of population distribution involved. Experience with various kinds of biological weight distributions, for example, shows that they tend

73

to possess a distribution very much like that shown in Fig. 2 of Chapter 2. Similarly, experience with distributions of various linear measurements, such as stature, foot length, and piston diameters shows that these variables possess distributions very much like that shown in Fig. 3 of Chapter 2. Thus, by combining experience and the information provided by the sample, one can usually postulate the general nature of the population distribution. This postulation leads to what is known as probability, or theoretical, distributions.

A probability distribution is a mathematical model for the actual frequency distribution. The probability models encountered in Chapter 3 in connection with games of chance are examples of probability distributions for certain discrete variables. Thus the postulation that each of the 36 possible outcomes in rolling two dice will occur equally often in the long run yields a probability distribution for the two dice. For the experiment of weighing 120 dormitory students, the model might be a continuous distribution, such as the bell-shaped distribution discussed in Chapter 2 in connection with the interpretation of the standard deviation. Models for continuous variables are more difficult to explain than those for discrete variables; therefore, discrete variables are considered first.

In discussing sample distributions and their theoretical counterparts, it is customary to call a sample distribution an *empirical* distribution. The relationship between an empirical distribution and its corresponding probability distribution has already been considered for a discrete variable in the illustration that produced Table 1 and Fig. 1 of Chapter 2. A sample of 700 was obtained by taking 700 drawings, with the drawn card replaced each time, from a set of cards consisting of 3 aces, 2 twos, and 1 three. From the results of the sampling, the relative

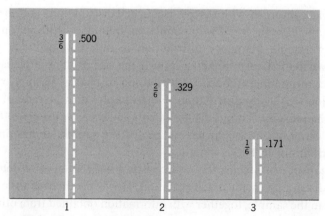

FIGURE 1.   A theoretical distribution and its empirical approximation.

frequencies for each type of outcome were calculated and, when expressed in decimal form, were found to be .500, .329, and .171, correct to three decimals. The corresponding probability distribution here is given by the three probabilities $\frac{3}{6}, \frac{2}{6}$, and $\frac{1}{6}$. A comparison of this probability distribution and its empirical approximation is shown in Fig. 1 as a pair of line charts, with the solid line representing theory and the broken line representing the sample. The approximation is so good here that it is not possible to detect any differences geometrically.

2. RANDOM VARIABLES

In experiments of the repetitive type for which a probability model is to be constructed one is usually interested only in a particular property of the outcome of the experiment. For example, in rolling two dice interest usually centers on the total number of points showing because that is all that really matters in the game of craps. Similarly, in taking samples of college students, interest might center on how many hours per week a student studies, or on a student's grade point average.

Just as in Chapter 2, the quantity chosen for study in such experiments will be denoted by the letter x. Thus, in the preceding illustrations, x might represent the sum of the points on two dice, or the number of hours per week a student studies, or a student's grade point average. In connection with probability distributions, the variable x is called a *random variable*.

For the purpose of seeing how a random variable is introduced in a simple experiment, return to the sample space for the coin-tossing experiment which is given in Fig. 1 of Chapter 3. If x is used to denote the total number of heads obtained, then each point of that sample space will possess the value of x shown directly above the corresponding point in Fig. 2. It will be observed that the random variable x can assume any one of the values 0, 1, 2, or 3, but no other values.

As another illustration, let x denote the total number of points obtained in rolling two honest dice. The sample space for this experiment is given by Table 1 of Chapter 3. This sample space has been duplicated in Fig. 3 but with the

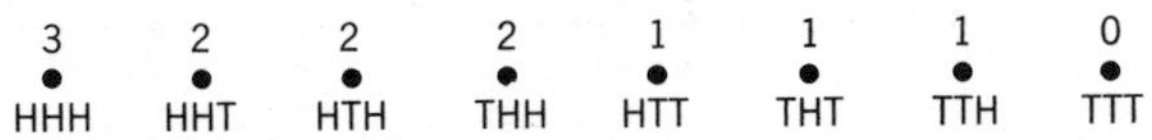

3	2	2	2	1	1	1	0
•	•	•	•	•	•	•	•
HHH	HHT	HTH	THH	HTT	THT	TTH	TTT

FIGURE 2. The values of a random variable for a coin experiment.

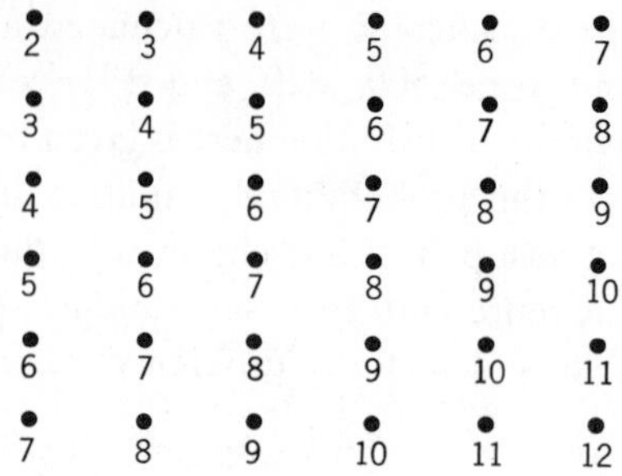

•	•	•	•	•	•
2	3	4	5	6	7
•	•	•	•	•	•
3	4	5	6	7	8
•	•	•	•	•	•
4	5	6	7	8	9
•	•	•	•	•	•
5	6	7	8	9	10
•	•	•	•	•	•
6	7	8	9	10	11
•	•	•	•	•	•
7	8	9	10	11	12

FIGURE 3. The values of a random variable for a dice experiment.

omission of the labels attached to the points. The numbers attached to the points are the values of the random variable x for this experiment. It will be observed that this random variable x can assume any one of the values 2, 3,..., 12.

In each of the preceding illustrations it will be observed that the value of the random variable x depends only on the particular sample point chosen. This means that x is a function of the sample points of the sample space. A formal definition is the following one:

Definition. A random variable is a numerical valued function defined on a sample space.

The word random, or chance, is used to designate variables of this type to point out that the value such a variable assumes in an experiment depends on the outcome of the experiment, which in turn depends on chance.

Now that interest is being centered on the values of a random variable for an experiment rather than on all the possible outcomes, a new simpler sample space can be constructed for the experiment, which can be substituted for the original sample space. Thus, in the coin-tossing experiment the only events of interest are the composite events given by $x = 0, 1, 2,$ and 3. But it follows from Fig. 2 and definition (1) of Chapter 3 that the probabilities for these composite events are given by $P\{0\} = \frac{1}{8}$, $P\{1\} = \frac{3}{8}$, $P\{2\} = \frac{3}{8}$, and $P\{3\} = \frac{1}{8}$, respectively. These composite events can now be treated as simple events in a new sample

FIGURE 4. Sample space for a coin-tossing random variable.

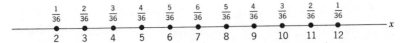

FIGURE 5. Sample space for a dice-rolling random variable.

space of four points with each point associated with a value of the random variable x. The probabilities just calculated for those composite events are then assigned to the four points of the new sample space. Figure 4 shows this new sample space with its associated probabilities.

In a similar manner, a new sample space for the random variable representing the total number of points showing on two dice can be constructed by means of Fig. 3 and definition (1) of Chapter 3. Here the composite events are those corresponding to the random variable x assuming the values $2, 3, \ldots, 12$. The probabilities for those composite events are readily calculated by means of definition (1) of Chapter 3 to be $\frac{1}{36}, \frac{2}{36}, \frac{3}{36}, \frac{4}{36}, \frac{5}{36}, \frac{6}{36}, \frac{5}{36}, \frac{4}{36}, \frac{3}{36}, \frac{2}{36}$, and $\frac{1}{36}$, respectively. As a result a new sample space consisting of eleven points based on the values of the random variable x can be constructed as shown in Fig. 5.

After a random variable and its corresponding sample space have been introduced and the probabilities to be assigned to those sample points calculated, the desired probability distribution for the random variable has been determined. The distribution of a random variable x is always understood to be its probability distribution and not its empirical distribution. Such a distribution consists of the values that x can assume and the probabilities associated with those values. Graphs of the distributions for the random variables of Figs. 4 and 5 are given in Figs. 6 and 7.

A probability distribution that has been constructed in the foregoing manner

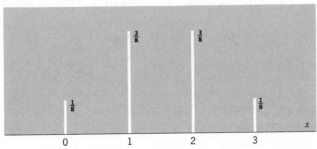

FIGURE 6. Distribution for a coin-tossing random variable.

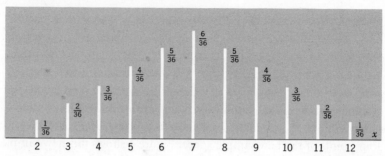

FIGURE 7.   Distribution for a random variable related to dice-rolling.

is to be considered as a mathematical model for a corresponding empirical distribution. Conversely, an empirical distribution of a variable x is to be considered as an approximation to a probability distribution for x.

3.   PROPERTIES OF PROBABILITY DISTRIBUTIONS

The discussion of empirical distributions in Chapter 2 began with a geometrical representation by means of histograms and then it proceeded to a partial arithmetic representation by means of the mean and standard deviation of the distribution. The same procedure will be followed for the probability distributions that are to be used as mathematical models for empirical distributions. The only essential difference in the calculations for a probability distribution as contrasted to those for an empirical distribution is that one uses probabilities in place of observed relative frequencies. Since a probability distribution corresponds to an empirical distribution for classified data, the earlier formulas for classified data will be used.

The formula for the mean given by (4), Chapter 2, can be written in the form

$$\bar{x} = \sum_{i=1}^{k} x_i \frac{f_i}{n}.$$

The corresponding theoretical mean, which is denoted by the Greek letter μ, is obtained by replacing the sample relative frequency f_i/n corresponding to the value x_i by the probability $P\{x_i\}$; hence

(1) $$\mu = \sum_{i=1}^{k} x_i P\{x_i\}.$$

If the sample extracted from a population is made increasingly large, it is to be expected that the relative frequency, f_i/n, of obtaining the value x_i will approach a fixed value. If so, this fixed value should be the probability $P\{x_i\}$. Thus, μ may be thought of as the value that $\bar{x}$ would be expected to approach as the sample size n becomes increasingly large. Table 1 and Fig. 1 of Chapter 2 illustrate how relative frequencies approach probability values as the sample size increases.

In a similar manner, the formula for the variance of an empirical distribution in its unmodified form and given by (6), Chapter 2, can be written as

$$s^2 = \sum_{i=1}^{k} (x_i - \bar{x})^2 \frac{f_i}{n}.$$

The corresponding theoretical variance, which is denoted by σ^2, is obtained by replacing f_i/n by $P\{x_i\}$ and replacing $\bar{x}$ by μ; hence

(2)
$$\sigma^2 = \sum_{i=1}^{k} (x_i - \mu)^2 P\{x_i\}.$$

The theoretical standard deviation, σ, is the positive square root of σ^2. There is no problem here, as there was in defining s^2, about whether to divide by n or $n - 1$, because n is not involved in theoretical definitions. The value of σ^2 may be thought of as the value that s^2 would be expected to approach as the sample size n becomes increasingly large.

As an illustration of the calculations involved in formulas (1) and (2), the mean and variance of the distribution given by Fig. 6 will be computed. Since the possible values of x are $x_1 = 0$, $x_2 = 1$, $x_3 = 2$, and $x_4 = 3$, with corresponding probabilities $P\{x_1\} = \frac{1}{8}$, $P\{x_2\} = \frac{3}{8}$, $P\{x_3\} = \frac{3}{8}$, and $P\{x_4\} = \frac{1}{8}$, it follows from formula (1) that

(3)
$$\mu = 0 \cdot \frac{1}{8} + 1 \cdot \frac{3}{8} + 2 \cdot \frac{3}{8} + 3 \cdot \frac{1}{8} = \frac{3}{2} = 1.5.$$

Similar calculations using formula (2) will yield

(4)
$$\sigma^2 = \left(0 - \frac{3}{2}\right)^2 \cdot \frac{1}{8} + \left(1 - \frac{3}{2}\right)^2 \cdot \frac{3}{8} + \left(2 - \frac{3}{2}\right)^2 \cdot \frac{3}{8} + \left(3 - \frac{3}{2}\right)^2 \cdot \frac{1}{8}$$
$$= \frac{9}{4} \cdot \frac{1}{8} + \frac{1}{4} \cdot \frac{3}{8} + \frac{1}{4} \cdot \frac{3}{8} + \frac{9}{4} \cdot \frac{1}{8} = \frac{3}{4}.$$

The mean and variance of a probability distribution are also called the mean and variance of the random variable x whose probability distribution is used in the calculations. Thus, the values of μ and σ^2 just calculated may be called the mean and variance of the random variable x, where x denotes the number of heads obtained in tossing three coins.

It is sometimes more convenient to calculate σ^2 by means of the following formula which is easily derived by squaring out the parentheses and summing the individual terms in (2).

(5)
$$\sigma^2 = \sum_{i=1}^{k} x_i^2 P\{x_i\} - \mu^2.$$

4.   EXPECTED VALUE

The mean and variance of a random variable are but two of its useful properties. They can be treated as special cases of a more general property called the *expected value*. For the purpose of describing what the phrase "expected value" means, consider a game in which you toss three honest coins and receive one dollar for each head that shows. How much money should you expect to get if you were permitted to play this game once? From Fig. 6 the probabilities associated with getting 0, 1, 2, and 3 heads in tossing three honest coins are $\frac{1}{8}, \frac{3}{8}, \frac{3}{8},$ and $\frac{1}{8}$, respectively. Intuitively, therefore, you should expect to get 0 dollars $\frac{1}{8}$ of the time, 1 dollar $\frac{3}{8}$ of the time, 2 dollars $\frac{3}{8}$ of the time, and 3 dollars $\frac{1}{8}$ of the time, if the game were played a large number of times. You should therefore expect to average the amount

$$\$0 \cdot \frac{1}{8} + \$1 \cdot \frac{3}{8} + \$2 \cdot \frac{3}{8} + \$3 \cdot \frac{1}{8} = \$1.50.$$

This amount is what is commonly called the expected amount to be won if the game is played once. It will be observed from (3) that this is the same as the mean value μ. This example is an illustration of the general concept of expected value that follows.

Suppose the random variable x must assume one of the values $x_1, x_2, \ldots, x_k$ and that the probabilities associated with those values are $P\{x_1\}, P\{x_2\}, \ldots, P\{x_k\}$, where $\sum_{i=1}^{k} P\{x_i\} = 1$. Then the expected value of this random variable is defined to be the quantity

(6)
$$E[x] = \sum_{i=1}^{k} x_i P\{x_i\}.$$

In the preceding illustration the random variable x was the amount of money to be won in tossing three coins, the possible values were 0, 1, 2, and 3, and the corresponding probabilities were $\frac{1}{8}$, $\frac{3}{8}$, $\frac{3}{8}$, and $\frac{1}{8}$.

A comparison of formula (1) with formula (6) shows that $E[x]$ is nothing more than the mean μ of the random variable x. Since the expected value of a random variable is the same as its mean it would appear that there is no point in introducing this new terminology; however, expectation goes a step further. Suppose the preceding game is altered so that you win the amount $g(x_i)$ instead of x_i when the value x_i is obtained. For example, $g(x)$ might be chosen to be the function $g(x) = x^2$. Then the expected value of the game to you would be given by the formula

$$(7) \qquad E[g(x)] = \sum_{i=1}^{k} g(x_i)P\{x_i\}.$$

Thus, it is possible to talk about the expected value of any function of a random variable, rather than just of the random variable itself. As an illustration, let $g(x) = x^2$ for the preceding example, then

$$E[g(x)] = \$0 \cdot \frac{1}{8} + \$1 \cdot \frac{3}{8} + \$4 \cdot \frac{3}{8} + \$9 \cdot \frac{1}{8} = \$3.$$

In view of this result, you could expect to win \$3 if the payoffs are the squares of the number of heads obtained and you are allowed to play the game once.

From definition (7) it is easily shown that the expected value operator E possesses the following properties:

$$(8) \qquad E[g(x) + c] = E[g(x)] + c$$

and

$$(9) \qquad E[cg(x)] = cE[g(x)].$$

A third property that is more difficult to demonstrate is given by the formula

$$(10) \qquad E[g(x) + h(y)] = E[g(x)] + E[h(y)].$$

In this last formula x and y are any two random variables and g and h are any two functions of those variables.

For the purpose of illustrating the meanings of these formulas, suppose that $g(x) = x$, $h(y) = y^2$, $c = 2$, and that the earlier game to illustrate expectation is being played, where x and y^2 denote the amounts to be won in the preceding two games. Then formula (8) says that if \$2 is added to each prize of the first game, the expected amount to be won in playing it once will be \$2 more than before. Formula (9) says that if each prize of the first game is doubled, the expected amount to be won will be doubled. Formula (10) says that if the amounts to be won are combined into a single game one can expect to win the sum of the expected values for the separate games. All of these formulas are intuitively obvious when one realizes that they represent the corresponding properties for mean values.

As another illustration of how to calculate expected values, consider the game in which you toss a coin three times and then toss a pair of dice. You are paid $5 for each head that shows and in addition as many dollars as the sum of the points showing on the two dice. How much should you expect to win? Let x represent the number of heads and y the total number of points that show. Then you should expect to win the amount $E[x + y] = E[x] + E[y]$. From Fig. 6 it follows that

$$E[x] = 0 \cdot \frac{1}{8} + 5 \cdot \frac{3}{8} + 10 \cdot \frac{3}{8} + 15 \cdot \frac{1}{8} = \frac{60}{8} = 7.50$$

and from Fig. 7 it follows that

$$E[y] = 2 \cdot \frac{1}{36} + 3 \cdot \frac{2}{36} + 4 \cdot \frac{3}{36} + 5 \cdot \frac{4}{36} + 6 \cdot \frac{5}{36} + 7 \cdot \frac{6}{36}$$
$$+ 8 \cdot \frac{5}{36} + 9 \cdot \frac{4}{36} + 10 \cdot \frac{3}{36} + 11 \cdot \frac{2}{36} + 12 \cdot \frac{1}{36}$$
$$= \frac{252}{36} = 7.$$

Hence, you should expect to win $14.50 if you are permitted to play this game once.

The variance of a probability distribution is conveniently represented by an expected value in the same manner as is the mean. This is accomplished by choosing $g(x) = (x - \mu)^2$ in (7). Then

$$E[(x - \mu)^2] = \sum_{i=1}^{k} (x_i - \mu)^2 P\{x_i\}.$$

A comparison of this result with (2) shows that

$$\sigma^2 = E[(x - \mu)^2].$$

Thus, the mean and variance are seen to be special cases of the expected value of a function of a random variable.

The expected value concept and its three foregoing properties will be used in Chapter 6 to assist in studying sampling problems and they will be used in Chapter 10 as a tool in probability decision making.

5.	CONTINUOUS VARIABLES

This section is concerned with a discussion of continuous random variables and their distribution. Such variables have already been studied in Chapter 2. The essential distinction between a continuous variable and a discrete variable is that the former involves measuring, whereas the latter involves counting. The variables of Chapter 2 for which the techniques of classification of data were explained were continuous variables, whereas the variables of Chapter 3 for which the rules of probability were derived were discrete variables.

Histograms are always used to represent empirical distributions of continuous variables, whereas line charts are normally used for discrete variable distributions. In Chapter 2, however, histograms were used in Figs. 4 and 5 even though the variables there are discrete.

For the purpose of discussing how probability distributions arise in connection with continuous variables, consider a particular continuous variable, x, that represents the diameter of a steel rod obtained from the production line of a manufacturer. If the next 200 rods coming off the production line were measured, there would be 200 values of x with which to study the diameter variation of the production system. Classifying these 200 values of x and graphing the histogram would help to describe the distribution of diameters. The histogram of Fig. 8 illustrates the results of one experiment. If 400 rods had been measured, the resulting histogram would have been about twice as tall as that in Fig. 8. This growth in the height of a histogram as the sample size increases makes it difficult to compare histograms based on different size samples. The difficulty can be overcome by requiring that the area of the histogram always be equal to 1. This can be accomplished by choosing the proper heights for the rectangles that make up the histogram.

The advantage of using a histogram whose area is equal to 1 becomes apparent when one calculates the relative frequencies for different sets of x values. Thus, from Fig. 8 the proportion of steel rods in the sample of 200 that had a diameter in inches between .4705 and .4755 is given by the area of the rectangle for which those numbers are the boundaries. This area is, of course, $23/200 = .115$ units.

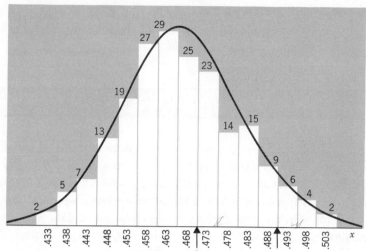

FIGURE 8. Distribution of the diameters of 200 steel rods, with a fitted normal curve.

Similarly, the proportion of steel rods in the sample that had a diameter in inches between .4705 and .4905 is given by the sum of the areas of the rectangles that begin at .4705 and end with .4905 and which are between the two arrows shown in Fig. 8. This area is given by $(23 + 14 + 15 + 9)/200 = .305$. Although it is not customary to use areas to calculate such relative frequencies, since they are readily obtained from the observed frequencies, it is important to realize that such relative frequencies can be represented by areas of parts of the histogram because theoretical relative frequencies will be calculated by means of areas of theoretical distributions.

With the foregoing choice of heights of rectangles to produce an area of 1, a histogram would be expected to approach a fixed histogram as the sample size is increased. For example, if a sample of 1000 steel rods were taken from the production line and the resulting histogram with area equal to 1 were graphed, one would not expect the histogram to change its shape much if additional samples were taken, because the histogram would already be an accurate estimate of the distribution of diameters for the production process. Now, if it is assumed that x can be measured as accurately as desired, so that the class interval can be made as small as desired, then the upper boundary of the histogram would be expected to settle down and approximate a smooth curve as the sample size is increased, provided the class interval is chosen very small. Such a curve is an idealization, or model, of the relative frequency with which different values of x would be expected to be obtained for runs of the actual experiment. The distribution given by such a curve is a theoretical distribution for the continuous variable x.

Figure 8 shows a curve that experience has indicated should conform closely to the kind of frequency distribution expected here and which therefore represents a theoretical distribution for that variable.

Since a curve that represents a theoretical distribution is thought of as the limiting form of the histogram under continuous sampling, it must possess the essential frequency properties of the histogram. Thus the area under a theoretical distribution curve must be equal to 1 because the area under the histogram is always kept equal to 1. Further, since the area of any rectangle of a histogram is equal to the relative frequency with which x occurred between the boundaries of the corresponding class interval, the area under the theoretical distribution curve between those same boundaries should represent the expected relative frequency of x occurring in that interval. If this same reasoning is extended to several neighboring intervals, it follows that the area under such a curve between any two values of x should represent the expected relative frequency of x occurring inside the interval determined by those two values of x.

As an illustration, consider once more the two values $x = .4705$ and $x = .4905$ that have been indicated by arrows in Fig. 8. The observed relative frequency for x occurring inside this interval is given by the area of the rectangles lying inside this interval and was found to be .305. If it is assumed that the curve sketched in Fig. 8 represents a proper theoretical distribution, then the area under this curve between $x = .4705$ and $x = .4905$ will represent the relative frequency with which x would be expected to occur inside this interval. This area was found, by methods to be explained later, to be equal to .335.

In Chapter 3 probability was defined for discrete variables in such a manner that it was interpreted as expected relative frequency for the event in question. Although the definition and rules of probability were restricted to discrete variables for ease of explanation, the rules also apply to continuous variable events. Consequently, in view of the discussion in the preceding paragraphs, it follows that a theoretical, or probability, distribution curve is a curve by means of which one can calculate the probability that x will lie inside any specified interval on the x axis. Thus, in the illustration related to Fig. 8, if it is assumed that the curve there represents the theoretical distribution, one can say that when a single sample value of x is taken the probability that the value of x will lie between .4705 and .4905 is .335. In symbols this would be written

$$P\{.4705 < x < .4905\} = .335.$$

This kind of statement is typical of probability statements for continuous variables; the events are usually concerned with the variable x lying inside some interval, or intervals, on the x axis. Such probabilities are given by areas under probability distribution curves.

As in the case for discrete variables, if one chooses a realistic model one can expect the probabilities calculated on the basis of that model to be close to the corresponding observed relative frequencies that are obtained when the experiment is repeated a large number of times.

Just as for discrete variables, the limiting values of $\bar{x}$ and s^2 for an empirical distribution of a continuous variable x are denoted by the letters μ and σ^2. The calculation of μ and σ^2 for a continuous variable probability distribution such as that shown in Fig. 8 requires a knowledge of calculus and therefore cannot be considered here.

6. ADDITIONAL ILLUSTRATIONS

1. A punch board has six punches left, and it is known that there are two $1 prizes left and four blanks. Let x denote the amount of money in dollars to be won. (a) 2 If one punch is to be taken, find the distribution of x and graph it. (b) 2 If two punches are to be taken, find the distribution of x and graph it. (c) 3 Calculate the mean and standard deviation of the distribution in part (a). The solutions follow.

(a) $\quad x = 0, 1$
$\quad\quad P\{x\} = \frac{4}{6}, \frac{2}{6}$

(b) $\quad x = 0 \quad, \quad 1 \quad, \quad 2$
$\quad\quad P\{x\} = \frac{4}{6} \cdot \frac{3}{5}, \ 2 \cdot \frac{2}{6} \cdot \frac{4}{5}, \ \frac{2}{6} \cdot \frac{1}{5}$

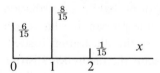

(c) $\mu = 0 \cdot \dfrac{4}{6} + 1 \cdot \dfrac{2}{6} = \dfrac{1}{3}$

$\sigma^2 = \left(0 - \dfrac{1}{3}\right)^2 \cdot \dfrac{4}{6} + \left(1 - \dfrac{1}{3}\right)^2 \cdot \dfrac{2}{6} = \dfrac{2}{9}, \sigma = \dfrac{\sqrt{2}}{3} = .47.$

If formula (5) had been used, the calculations would have been

$$\sigma^2 = 0^2 \cdot \frac{4}{6} + 1^2 \cdot \frac{2}{6} - \left(\frac{1}{3}\right)^2 = \frac{2}{6} - \frac{1}{9} = \frac{2}{9}.$$

2.   A box contains the following nine cards: the three, four, and five of spades, the three and four of clubs, the three and four of hearts, and the four and five of diamonds. (a) 2 If one card is to be drawn from the box and x is the random variable representing the number of black cards that will be obtained, find the distribution of x and graph it. (b) 2 If one card is to be drawn and x represents the number on the card, find the distribution of x and graph it. (c) 2 If two cards are to be drawn with the first card being replaced before the second drawing and x represents the number of black cards that will be obtained, find the distribution of x and graph it. (d) 2 Work part (c) if the first card is not replaced. (e) 2 If two cards are to be drawn without replacement and x represents the sum of the two numbers obtained, find the distribution of x and graph it. (f) 3 Calculate the mean and standard deviation of the distribution in (c). The solutions follow.

(a)

(b)

(c)

Outcome	RR	RB	BR	BB
Probability	$\left(\frac{4}{9}\right)^2$	$\left(\frac{4}{9}\right)\left(\frac{5}{9}\right)$	$\left(\frac{5}{9}\right)\left(\frac{4}{9}\right)$	$\left(\frac{5}{9}\right)^2$
x	0	1	1	2

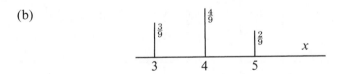

(d)

Outcome	RR	RB	BR	BB
Probability	$\frac{4}{9} \cdot \frac{3}{8}$	$\frac{4}{9} \cdot \frac{5}{8}$	$\frac{5}{9} \cdot \frac{4}{8}$	$\frac{5}{9} \cdot \frac{4}{8}$
x	0	1	1	2

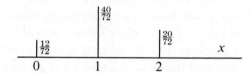

(e) Outcomes: 33 43 53   Values of x: 6 7 8
             34 44 54   7 8 9
             35 45 55   8 9 10

Probabilities: $\frac{3}{9} \cdot \frac{2}{8}$ $\frac{4}{9} \cdot \frac{3}{8}$ $\frac{2}{9} \cdot \frac{3}{8}$
               $\frac{3}{9} \cdot \frac{4}{8}$ $\frac{4}{9} \cdot \frac{3}{8}$ $\frac{2}{9} \cdot \frac{4}{8}$
               $\frac{3}{9} \cdot \frac{2}{8}$ $\frac{4}{9} \cdot \frac{2}{8}$ $\frac{2}{9} \cdot \frac{1}{8}$

x	6	7	8	9	10
$P\{x\}$	$\frac{6}{72}$	$\frac{24}{72}$	$\frac{24}{72}$	$\frac{16}{72}$	$\frac{2}{72}$

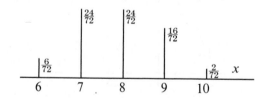

(f)

x	$81P\{x\}$	$81xP\{x\}$	$81x^2P\{x\}$
0	16	0	0
1	40	40	40
2	25	50	100
		90	140

$$\mu = \frac{90}{81} = \frac{10}{9}$$

Using formula (5),

$$\sigma^2 = \frac{140}{81} - \left(\frac{10}{9}\right)^2 = \frac{40}{81}$$

$$\sigma = \frac{\sqrt{40}}{9}$$

3. An individual who owns the ice cream concession at a sporting event can expect to net $600 on the sale of ice cream if the day is sunny, but only $300 if it is cloudy, and $100 if it rains. The respective probabilities for those events are .6, .3, and .1. (a) 4 What is his expected profit? (b) 4 If he takes out $400 worth of insurance against rain and the insurance costs him $90, what is his expected profit?

(a) $E = 600(.6) + 300(.3) + 100(.1) = 460$.

(b) If it rains he will realize both $100 and the insurance; hence

$$E = 600(.6) + 300(.3) + 500(.1) - 90 = 410.$$

EXERCISES

Section 2

1. Toss 3 coins simultaneously and record the number of heads obtained. Perform this experiment 100 times and then compare your experimental relative frequencies with those given by theory in Fig. 6.

2. Roll a die twice, or roll 2 dice simultaneously, and record the number of points obtained. Perform this experiment 100 times and compare your experimental relative frequencies with those given by theory in Fig. 7.

3. A box contains three cards consisting of the two, three, and four of hearts. Two cards are drawn from the box, with the first card drawn returned to the box before the second drawing. Let x denote the sum of the numbers obtained on the two cards. Use the enumeration of events technique to derive the distribution of the random variable x.

4. Work problem 3 under the assumption that the first card is not returned to the box.

5. Calculate the values of μ and σ for the distribution obtained in problem 3.

6. Calculate the values of μ and σ for the distribution obtained in problem 4.

7. Calculate the values of μ and σ for the distribution of random digits, that is, for $f(x) = \frac{1}{10}$, $x = 0, 1, 2, \ldots, 9$.

8. Calculate the values of μ and σ for the distribution given by $f(-1) = \frac{3}{8}$, $f(0) = \frac{2}{8}$, $f(1) = \frac{3}{8}$. What percentage of the distribution is included in the interval $(\mu - \sigma, \mu + \sigma)$?

Section 4

9. You toss a coin three times. If you get at least two heads you are permitted to roll a die and you receive in dollars the number of points that show. What can you expect to win at this game if you are permitted to play it once?

10. You toss a coin three times. If you get less than three heads you receive in dollars the number of heads obtained. If you get three heads you are permitted to roll a die and you receive in dollars the number of points showing plus a two dollar bonus. What is your expected value in this game?

11. Which game would you choose if given a choice: toss two dice and receive in dollars the sum of the points showing or toss four coins and receive in dollars double the number of heads obtained?

12. A sports promoter is contemplating taking out rain insurance for an athletic event he is sponsoring. If it does not rain he can expect to net $10,000 on the event, but only $2000 if it does. A $7000 insurance policy will cost him $3000. Determine the probability of rain, p, such that his expectation will be the same whether or not he takes out insurance.

13. A car owner wishes to sell his car and is contemplating spending $50 to advertise it. If the probability is .5 that he will sell it at his stipulated price of $750 without advertising and is .9 of doing so if he does advertise, should he advertise? Assume that if he does not sell it for $750 he will let a friend have it for $650.

12. $10,000 X + (1-X)2000 = 10,000 + (1-X)(9000) - 3000$

$10,000 X + 2000 - 2000X = 10,000 + 9000 - 9000X - 3000$

$7000X = 4000$

$X = 4/7$

$P(rain) = 3/7$

13. $.5(750) + .5(650)$

$.9(700) + .1(600)$

CHAPTER 5

Some Special Probability Distributions

In this chapter two of the most useful probability distributions for solving statistical inference problems will be introduced. These distributions, one for a discrete variable and the other for a continuous variable, are undoubtedly the most commonly used distributions in statistical practice.

1. BINOMIAL DISTRIBUTION

Consider an experiment in which each of the possible outcomes can be classified as resulting or not resulting in the occurrence of an event A. If it resulted in the occurrence of A it will be classified as a success, otherwise as a failure. The word success is used here as a convenient way of describing the occurrence of an event but it does not imply that the occurrence of the event is necessarily desired. The experiment will be repeated a number of times, this number being denoted by the letter n. A random variable x will be introduced which represents the total number of successes, that is, occurrences of A, that were obtained in the n repeti-

91

tions of the experiment. A random variable of this type is called a *binomial variable*.

The coin-tossing experiment that has been used so frequently can also be used here to give an example of a binomial variable and its distribution. Let that experiment be changed to consist of tossing the coin once instead of three times. Success will be defined as getting a head. The experiment will be repeated three times; hence $n = 3$ here. The random variable x will then represent the number of heads obtained in the three tosses, just as it did in section 2, Chapter 4. The distribution of this binomial random variable is therefore given by Fig. 6, Chapter 4. It is reproduced here in Fig. 1.

The experiment of rolling two dice with x defined as the sum of the points on the two dice does not produce a binomial variable because one cannot classify each roll of a die as producing either a success or a failure and then have x represent the sum of the successes.

The experiment of drawing a ball from a box consisting of three red, two black, and one green ball is also not an experiment that leads to a binomial variable in its present form because there are three possible outcomes here rather than two. However, if one were interested only in knowing whether a red ball will be obtained, then the problem becomes a binomial distribution problem. The variable x would then represent the number of red balls obtained in performing the experiment n times. It is always understood in binomial variable problems that replacements are made before the next experiment is begun when it consists of drawing objects from containers. The repetitions of the experiment must be repetitions of the original experiment in every sense.

Now look at a slightly more complicated example of a binomial random variable and how its distribution is obtained. The basic experiment will consist of rolling a die once and success will be defined as getting an ace (one-spot).

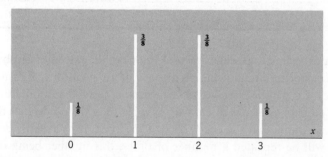

FIGURE 1. Distribution for a coin-tossing random variable.

The experiment will be performed three times, so that $n = 3$, and the random variable x will represent the number of aces obtained in the three rolls. To obtain the distribution of this random variable one can proceed in the same manner as for the coin-tossing experiment of Fig. 1. This consists of looking at the original sample space for the complete experiment and then reducing it to a new sample space for the random variable x by applying the definition of probability of events to the original sample space. Now for each roll of the die it is merely necessary to record whether a success or a failure occurred, where success corresponds to an ace showing. If S and F are used to represent success and failure, then there are eight points in the sample space, just as there were for three tosses of a coin in which either an H or a T must occur at each toss. These points have been represented in Table 1 by means of the letters S and F to indicate the various possible outcomes. The corresponding values of the random variable x have also been displayed in Table 1.

TABLE 1

Outcome	SSS	SSF	SFS	FSS	SFF	FSF	FFS	FFF
Value of x	3	2	2	2	1	1	1	0

It will be observed that this table is precisely the same as for the problem of tossing a coin three times and which is shown in Fig. 2, Chapter 4. However, the calculation of the probabilities for the various values of x is considerably different. The probabilities for these eight possible outcomes are not equal as was the case for the coin problem. The probabilities here will be calculated by using the multiplication rule of probability. Because of the independence of the three rolls of the die and the fact that the probability of a success (getting an ace) is now $\frac{1}{6}$, it follows, for example, that

$$P\{\text{SFS}\} = \frac{1}{6} \cdot \frac{5}{6} \cdot \frac{1}{6} = \left(\frac{1}{6}\right)^2\left(\frac{5}{6}\right).$$

Calculations similar to this will yield probabilities for each of the eight possible outcomes. The results of such calculations are shown in Table 2.

TABLE 2

Outcome	SSS	SSF	SFS	FSS	SFF	FSF	FFS	FFF
Probability	$\left(\frac{1}{6}\right)^3$	$\left(\frac{1}{6}\right)^2\left(\frac{5}{6}\right)$	$\left(\frac{1}{6}\right)^2\left(\frac{5}{6}\right)$	$\left(\frac{1}{6}\right)^2\left(\frac{5}{6}\right)$	$\left(\frac{1}{6}\right)\left(\frac{5}{6}\right)^2$	$\left(\frac{1}{6}\right)\left(\frac{5}{6}\right)^2$	$\left(\frac{1}{6}\right)\left(\frac{5}{6}\right)^2$	$\left(\frac{5}{6}\right)^3$

This sample space of eight points with its assigned probabilities can now be reduced to the sample space for the random variable x by calculating the probabilities of the composite events corresponding to the various values of the random variable. This is done by comparing Tables 1 and 2. Thus, the event $x = 2$ comprises the simple events SSF, SFS, FSS; therefore its probability is obtained by adding the probabilities associated with those three points. The results of such calculations are shown in Table 3. This table gives the distribution of the desired random variable. Its graph is shown in Fig. 2, in which the probabilities have been expressed in decimal form to two decimals.

TABLE 3

x	0	1	2	3
$P\{x\}$	$(\frac{5}{6})^3$	$3(\frac{1}{6})(\frac{5}{6})^2$	$3(\frac{1}{6})^2(\frac{5}{6})$	$(\frac{1}{6})^3$

In view of these results, it is clear that one should not expect to get three aces when rolling three dice. Such a result will occur in the long run about once in $6^3 = 216$ experiments. It is also clear that one should not accept a wager based on even money that one will get at least one ace in rolling three dice. Such a result will occur about 42 per cent of the time; therefore if you wish to get the better of your naïve friends, wager them that they will not get at least one ace when rolling three dice.

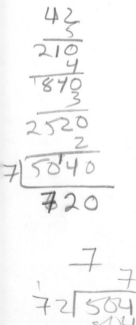

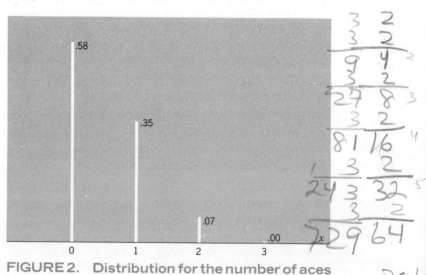

FIGURE 2.   Distribution for the number of aces in rolling a die three times.

The technique employed in the two preceding illustrations can be used on any particular binomial problem that arises. One first constructs the sample space for the complete experiment. If there are, say, five repetitions of the basic success or failure experiment, there will be $2^5 = 32$ points in the sample space. The probability for each point is then calculated. Next, the proper value of the random variable x is associated with each point of the sample space. Then by summing the probabilities of those points that correspond to a particular value of x, the probabilities for the various values of x are obtained. These probabilities give the desired probability distribution of the binomial random variable x.

Because of the difficulty of carrying through the preceding computations each time a binomial problem arises it is convenient to have a formula that is applicable to all such problems. For this purpose consider a general binomial problem. An experiment is to be performed for which the outcome can always be classified as either a success or a failure. The probability that it will produce a success is assumed given and is denoted by the letter p. The corresponding probability of a failure is denoted by q; hence $p + q = 1$. The experiment is to be performed n times. The number of successes that will be obtained in the n repetitions of the experiment is denoted by the letter x. The problem then is to calculate the probabilities for the various possible values of the random variable x. As stated before, these computations can be carried out systematically for any given problem and they can also be done here for this general problem if the knowledge of optional section 7 of Chapter 3 is available. This is done in section 5 of this chapter for the benefit of those who have had time to study that earlier material; therefore only the results of those computations will be given here. The symbol $P\{x\}$ will be used to denote the probability for a typical value of this general binomial variable. These probabilities, which define what is known as the binomial distribution, are given by the following formula.

(1)   *Binomial Distribution.* $P\{x\} = \dfrac{n!}{x!(n-x)!} p^x q^{n-x}.$

As explained in section 7 of Chapter 3, the symbol $n!$ is read "n factorial." It denotes the product of all the positive integers from 1 through n. Thus, $5! = 1 \cdot 2 \cdot 3 \cdot 4 \cdot 5$. The value of $0!$ is defined to be 1 rather than 0, as might have been assumed. For large values of n and x the computations involved in evaluating the quantity $n!/x!(n-x)!$ become rather heavy; consequently a table of values of this quantity for various values of n and x has been made available in the appendix as Table III. This table handles values of n from 2 through 20, and for x from 2 through $n/2$ or more. It is not necessary to list the values for larger values of x because if x is replaced by $n - x$ in this quantity the same value will result; consequently one uses this device to handle values of x larger than

$n/2$. Although this table enables one to write down binomial probabilities rather quickly it will not be used in the illustrative problems that follow because it seems desirable for the student to become acquainted with formula (1) by carrying out the required computations when n is fairly small.

It is customary to speak of the n experiments as n independent trials of an experiment for which p is the probability of success in a single trial. In this language $P\{x\}$ is the probability of obtaining x successes in n independent trials of an experiment for which p is the probability of success in a single trial.

The two frequency distributions displayed in Figs. 1 and 2 are special cases of the general binomial distribution given by (1). For the coin problem, $n = 3$ and $p = q = \frac{1}{2}$. For the die problem, $n = 3, p = \frac{1}{6}$, and $q = \frac{5}{6}$. The values of $P\{x\}$ given in Fig. 1 and Fig. 2 should be checked by means of formula (1) for the purpose of becoming familiar with its use. As an illustration of such a check, if the value of $P\{3\}$ is desired for the die problem, substitution of the proper values into (1) will yield

$$ P\{3\} = \frac{3!}{3!\,0!}\left(\frac{1}{6}\right)^{3}\left(\frac{5}{6}\right)^{0}. $$

Since $0! = 1$ and since by algebra any number to the 0 power equals 1 it follows that $(\frac{5}{6})^0 = 1$, and hence that

$$ P\{3\} = \left(\frac{1}{6}\right)^{3}. $$

Although the problems used to introduce the binomial distribution were related to games of chance, the binomial distribution is very useful for solving certain types of practical problems. Such problems are solved in the next few chapters; meanwhile, a few simple problems which require only easy computations with formula (1) will be discussed.

The probability that parents with a certain type of blue-brown eyes will have a child with blue eyes is $\frac{1}{4}$. If there are six children in the family, what is the probability that at least half of them will have blue eyes? To solve this problem the six children in the family will be treated as six independent trials of an experiment for which the probability of success in a single trial is $\frac{1}{4}$. Thus $n = 6$ and $p = \frac{1}{4}$ here. It is necessary to calculate $P\{3\}$, $P\{4\}$, $P\{5\}$, and $P\{6\}$ and sum because these probabilities correspond to the mutually exclusive ways in which the desired event can occur. By formula (1),

$$ P\{3\} = \frac{6!}{3!\,3!}\left(\frac{1}{4}\right)^{3}\left(\frac{3}{4}\right)^{3} = \frac{540}{4096}, $$

$$P\{4\} = \frac{6!}{4!\,2!}\left(\frac{1}{4}\right)^4\left(\frac{3}{4}\right)^2 = \frac{135}{4096},$$

$$P\{5\} = \frac{6!}{5!\,1!}\left(\frac{1}{4}\right)^5\left(\frac{3}{4}\right)^1 = \frac{18}{4096},$$

$$P\{6\} = \frac{6!}{6!\,0!}\left(\frac{1}{4}\right)^6\left(\frac{3}{4}\right)^0 = \frac{1}{4096}.$$

The probability of getting at least three successes is obtained by adding these probabilities; consequently, writing $x \geq 3$ to represent at least three successes, one obtains

$$P\{x \geq 3\} = \frac{694}{4096} = .169.$$

This result shows that there is a very small chance that a family such as this will have so many blue-eyed children. In only about 17 of 100 such families will at least half the children be blue-eyed.

A manufacturer of certain parts for automobiles guarantees that a box of his parts will contain at most two defective items. If the box holds 20 parts and experience has shown that his manufacturing process produces 2 per cent defective items, what is the probability that a box of his parts will satisfy the guarantee? This problem can be considered as a binomial distribution problem for which $n = 20$ and $p = .02$. A box will satisfy the guarantee if the number of defective parts is 0, 1, or 2. By means of formula (1) the probabilities of these three events are given by

$$P\{0\} = \frac{20!}{0!\,20!}(.02)^0(.98)^{20} = (.98)^{20} = .668,$$

$$P\{1\} = \frac{20!}{1!\,19!}(.02)^1(.98)^{19} = 20(.02)(.98)^{19} = .273,$$

$$P\{2\} = \frac{20!}{2!\,18!}(.02)^2(.98)^{18} = 190(.02)^2(.98)^{18} = .053.$$

The calculations here were made with the aid of logarithms. Since these are mutually exclusive events, the probability that there will be at most two defective parts, written $x \leq 2$, is the sum of these probabilities; hence the desired answer is

$$P\{x \leq 2\} = .994.$$

This result shows that the manufacturer's guarantee will almost always be satisfied.

As a final illustration consider the following problem concerning whether it pays to guess on an examination. Suppose an examination consists of a large number of questions of the multiple choice type, with each question having five possible answers but only one of the five being the correct answer. If a student receives 3 points for each correct answer and -1 point for each incorrect answer and if on each of ten of the questions his probability of guessing the correct answer is only $\frac{1}{3}$, what is his probability of obtaining a total positive score on those ten questions?

If x denotes the number of questions answered correctly, then a positive score will result if $3x > 10 - x$ because the left side of this inequality gives the total number of positive points scored and the right side the total number of penalty points. This inequality will be satisfied if $x > \frac{10}{4}$, which implies that at least three correct answers must be obtained to realize a positive score. The desired probability is therefore given by

$$P\{x \geq 3\} = 1 - \sum_{x=0}^{2} \frac{10!}{x!(10-x)!}\left(\frac{1}{3}\right)^{x}\left(\frac{2}{3}\right)^{10-x}$$

$$= 1 - \left[\left(\frac{2}{3}\right)^{10} + 10\left(\frac{1}{3}\right)\left(\frac{2}{3}\right)^{9} + 45\left(\frac{1}{3}\right)^{2}\left(\frac{2}{3}\right)^{8}\right]$$

$$= .70.$$

Thus, he has an excellent chance of gaining points if his probability of guessing a correct answer is as high as $\frac{1}{3}$. If he knew nothing about the material and selected one of the five alternatives by chance, his probability would, of course, be only $\frac{1}{5}$ for each question. It is assumed here, however, that he knows enough about the subject to be able to discard two of the five possibilities as being obviously incorrect and make a guess among the other three. If he had no such knowledge, so that his probability would be $\frac{1}{5}$, then similar calculations would show that it does not pay to guess.

2.   BINOMIAL DISTRIBUTION PROPERTIES

Since general formulas were made available in Chapter 4 for calculating the mean and variance of a discrete random variable, those formulas may be used to obtain the mean and variance of any binomial variable. The calculations that produced the values given in (3) and (4), Chapter 4, are those needed for the binomial variable for which $p = \frac{1}{2}$ and $n = 3$.

For the purpose of becoming more familiar with such computations, consider the problem of calculating the mean and variance of the binomial variable given by Table 3 and whose graph is shown in Fig. 2. Here $p = \frac{1}{6}$, $n = 3$, and the possible values of x are $x_1 = 0$, $x_2 = 1$, $x_3 = 2$, and $x_4 = 3$. The corresponding values of the $P\{x_i\}$ are given in Table 3. Application of formula (1), Chapter 4, then gives

$$\mu = 0 \cdot \left(\frac{5}{6}\right)^3 + 1 \cdot 3\left(\frac{5}{6}\right)^2\left(\frac{1}{6}\right) + 2 \cdot 3\left(\frac{5}{6}\right)\left(\frac{1}{6}\right)^2 + 3 \cdot \left(\frac{1}{6}\right)^3$$

$$= \frac{108}{216} = \frac{1}{2}.$$

Using this value of μ and applying formula (2), Chapter 4, will give

$$\sigma^2 = \frac{1}{4} \cdot \left(\frac{5}{6}\right)^3 + \frac{1}{4} \cdot 3\left(\frac{5}{6}\right)^2\left(\frac{1}{6}\right) + \frac{9}{4} \cdot 3\left(\frac{5}{6}\right)\left(\frac{1}{6}\right)^2 + \frac{25}{4} \cdot \left(\frac{1}{6}\right)^3$$

$$= \frac{90}{216} = \frac{5}{12}.$$

As a result, $\sigma = \sqrt{\frac{5}{12}}$.

It is possible to employ some algebraic tricks to carry out similar calculations for the general binomial distribution given by (1). Such calculations yield formulas for the mean and standard deviation that can be used for all binomial problems. Since such calculations are rather complicated, they will not be carried out here; however, the resulting formulas will be given and used for solving binomial problems. These formulas, which are very simple, are

(2)
$$\mu = np$$
$$\sigma = \sqrt{npq}.$$

The advantage of having such neat formulas becomes apparent when one applies them to the computational problem that was just completed. For the distribution of Table 3, $n = 3$ and $p = \frac{1}{6}$; therefore formulas (2) give

$$\mu = 3 \cdot \frac{1}{6} = \frac{1}{2}$$

$$\sigma = \sqrt{3 \cdot \frac{1}{6} \cdot \frac{5}{6}} = \sqrt{\frac{5}{12}}.$$

3.   NORMAL DISTRIBUTION

The binomial distribution of section 1 is the most useful probability distribution for discrete variables. The distribution that will be studied in this section is the most useful probability distribution for continuous variables.

The histogram in Fig. 8, Chapter 4, is typical of many distributions found in nature and industry. Such distributions are quite symmetrical, die out rather quickly at the tails, and possess a shape much like that of a bell. A theoretical distribution, which has proved very useful for distributions such as these and which will presently be seen as very important in other ways also, is called the *normal distribution*. The curve that has been sketched in Fig. 8, Chapter 4, is the graph of a particular normal distribution. The graph of a general normal distribution is given in Fig. 3. Although a normal distribution is defined by the equation of its curve, this equation is not used explicitly in subsequent chapters and therefore it is not written down. The curve itself can be thought of as defining the distribution.

You may have heard of the normal curve in connection with the distribution of grades in large sections of certain courses. Grading on the basis of the normal curve assumes that the distribution of the mental output of students should be similar to the distribution of many of their physical characteristics, which are known to be approximately normally distributed. An instructor using such a grading scheme also usually assumes that the students in his course have the proper prerequisites and study the expected number of hours. An optimist at

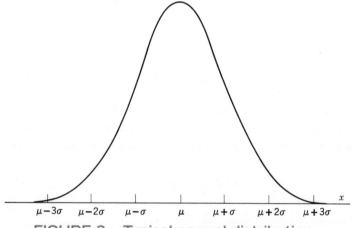

FIGURE 3.   Typical normal distribution.

heart! One difficulty with this system is that it assigns grades on a relative rather than an absolute basis. Thus an unusually gifted class that works hard will be assigned the same distribution of grades as a class of lazy louts. Of course, the answer to this criticism is that students from year to year differ very little, and therefore it is highly unlikely that a large class will be made up of either brilliant workers or stupid loafers. If you have acquired a dislike for the normal curve on the basis of the grades it may have given you, realize that the normal curve is really not to blame and be charitable enough to approach its study here without prejudice. It is an exceedingly useful curve in many fields of application quite removed from school problems.

It will be recalled from Chapter 2 that the mean of a distribution represents the point on the x axis at which a sheet of metal in the shape of the histogram of the distribution will balance on a knife-edge. This geometrical property of the mean makes it clear that when a histogram is symmetrical about a vertical axis the mean must be located at the symmetry point on the x axis. This is true for the limiting value of the mean as well when the size of the sample is increased indefinitely and the class interval is made increasingly small. This explains why the symmetry point on the x axis on Fig. 3 has been labeled μ.

Suppose, now, that the limiting form of the histogram for a frequency distribution, in the sense described earlier, is a normal curve. Then it can be shown by advanced mathematical methods that σ, the limiting value of s, has the following geometrical interpretation with respect to the limiting normal curve:

(3)   (a) The area under the normal curve between $\mu - \sigma$ and $\mu + \sigma$ is 68 per cent of the total area, to the nearest 1 per cent.

(b) The area under the normal curve between $\mu - 2\sigma$ and $\mu + 2\sigma$ is 95 per cent of the total area, to the nearest 1 per cent.

(c) The area under the normal curve between $\mu - 3\sigma$ and $\mu + 3\sigma$ is 99.7 per cent of the total area, to the nearest .1 per cent.

The axis in Fig. 3 has been marked off in units of σ, starting with the mean μ. It is clear from this sketch that there is almost no area under the curve beyond 3σ units from μ; however, the equation of the curve would show that the curve actually extends from $-\infty$ to $+\infty$.

The first two of the foregoing properties were used to give meaning to s in Chapter 2. Those two percentages were found there to be approximately correct for histograms whose shapes resemble a normal curve.

An interesting property of the normal curve is that its location and shape are completely determined by its values of μ and σ. The value of μ, of course, centers the curve, whereas the value of σ determines the extent of the spread. Since all

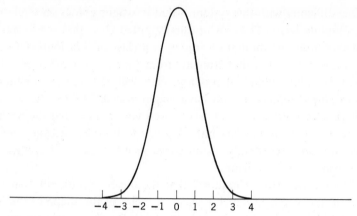

FIGURE 4.   A normal distribution with $\mu = 0$ and $\sigma = 1$.

normal curves representing theoretical frequency distributions have a total area of 1, as σ increases the curve must decrease in height and spread out. This is illustrated in Figs. 4 and 5 which give sketches of two normal curves with the same mean, namely 0, and standard deviations of 1 and 3, respectively. The fact that the shape of a normal curve is completely determined by its standard deviation enables one to reduce all normal curves to a standard one by a simple change of variable. For example, the curve of Fig. 4 can be made to look like the curve of Fig. 5 by changing the scale on the x axis so that one unit on the Fig. 4 axis represents three units on the Fig. 5 axis. This corresponds, roughly, to taking the curve of Fig. 4, treating it and the area beneath it as though it were made of rubber, and stretching it out to three times its natural length with its area preserved. Conversely, Fig. 5 could be made to go into Fig. 4 by compressing Fig. 5 to one-third its natural length. This assumes that the long tails of these curves are cut off and ignored. Since the simplest normal curve with which to work is the one that has its mean at 0 and whose standard deviation is equal to 1, other normal curves are usually reduced to this standard one when there is need for a reduction. Now to any point on the x axis of a normal curve there corresponds

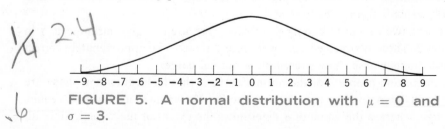

FIGURE 5.   A normal distribution with $\mu = 0$ and $\sigma = 3$.

a point on the x axis of the standard normal curve, and its value can be determined by stating how many standard deviations it is away from the mean point of the curve. Thus, the point $x = 6$ on Fig. 5 corresponds to the point $x = 2$ on the standard normal curve given by Fig. 4; therefore the value $x = 6$ can be obtained from Fig. 4 by stating that it is 2 standard deviations to the right of its mean 0. In general, if a point x on the axis of a normal curve with mean μ and standard deviation σ corresponds to a point z on the standard normal curve, then the point x is z standard deviations to the right of μ. The relationship between these corresponding points is therefore given by the formula

$$x = \mu + z\sigma. \quad \Leftarrow$$

Or, if z is expressed in terms of x,

(4)
$$z = \frac{x - \mu}{\sigma}.$$

This formula enables one to find the point z on the standard normal curve that corresponds to any point x on a nonstandard normal curve. Thus, the point $x = 4$ on Fig. 5 corresponds to the point $z = (4 - 0)/3 = 1\frac{1}{3}$ on Fig. 4. By this device of expressing all x values on a normal curve in terms of corresponding values on the standard normal curve, all normal curves can be reduced to a single standard one.

In the earlier discussions in Chapter 2, it was noted that the standard deviation, s, is unaffected by adding a constant to the values of a set of measurements and that it is multiplied by c if each of the measurements is multiplied by c. This same property will hold for a random variable x and its theoretical standard deviation σ. That is, the variable $x - c$ will have the same standard deviation as the variable x, but the standard deviation for cx will be c times as large as for x. In view of these properties the standard deviation of $(x - \mu)/\sigma$ will be $1/\sigma$ times the standard deviation of $x - \mu$, or of x; therefore if σ is the standard deviation of x, the standard deviation of $z = (x - \mu)/\sigma$ must be 1. Since the mean of x is μ, subtracting μ from x will give a variable $x - \mu$ whose mean is 0. The mean of the variable $z = (x - \mu)/\sigma$ is therefore also 0, because multiplying a variable by a constant multiplies the mean by that constant, and multiplying 0 by $1/\sigma$ still gives 0. Thus, the variable $z = (x - \mu)/\sigma$ will possess the mean 0 and the standard deviation 1. The change of variable given by formula (4) will therefore change any variable x to one with mean 0 and standard deviation 1. This is true whether the variable x is a normal variable or not. A variable x that has been changed to the variable z by means of formula (4) is said to be measured in *standard units* after the change has been made.

Table IV in the appendix is a table for finding the area under any part of the normal curve for the variable z, that is, for the normal curve that has mean 0 and standard deviation 1. The values of z in this table are given to two decimal places, with the second decimal place determining the column to use. As an illustration, suppose one wished to find the area under this standard normal curve from $z = 0$ to $z = 1.00$. The desired area is shown geometrically in Fig. 6. In Table IV one reads down the first column until the z value 1.0 is reached, then across to the entry in the column headed .00 to find .3413. This is the desired area. Table IV gives only areas from $z = 0$ to any specified positive value of z. If areas to the left of $z = 0$ are wanted, one must use symmetry and work with the corresponding right half of the curve.

Suppose now that one wishes to find the area under part of a normal curve with mean μ and standard deviation σ. For example, suppose one wishes to find the area from $x = \mu - \sigma$ to $x = \mu + \sigma$ in Fig. 3. It follows, by symmetry, that this area is twice the area from $x = \mu$ to $x = \mu + \sigma$, which is precisely the same as the corresponding area under the standard normal curve of Table IV that is sketched in Fig. 6. From formula (4), the values $x = \mu$ and $x = \mu + \sigma$ correspond to the values $z = 0$ and $z = 1$; consequently, the area from $x = \mu$ to $x = \mu + \sigma$ is the same as that under the standard normal curve from $z = 0$ to $z = 1$. Since this area was found to be .3413, the area from $x = \mu - \sigma$ to $x = \mu + \sigma$ is twice this number, or .6826. This, of course, is the number that gave rise to the normal distribution property (a) in (3).

As another illustration of the use of Table IV, suppose one wishes to find the area between $x = 220$ and $x = 280$ for x possessing a normal distribution with

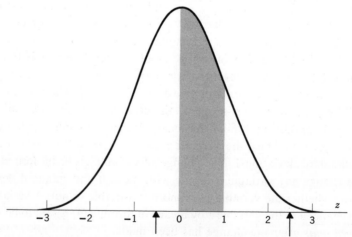

FIGURE 6.   Standard normal distribution.

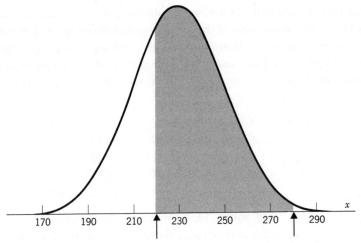

FIGURE 7. A particular normal distribution.

$\mu = 230$ and $\sigma = 20$. The desired area is shown in Fig. 7. First it is necessary to calculate the corresponding z values by means of (4). These are

$$z_1 = \frac{220 - 230}{20} = -.50$$

and

$$z_2 = \frac{280 - 230}{20} = 2.50.$$

These two z values are indicated by means of vertical arrows on Fig. 6. Now, the desired area is given by the area from $-.50$ to 2.50 under the standard normal curve. From Table IV, the area from $z = 0$ to $z = 2.50$ is $.4938$. By symmetry, the area from $z = -.50$ to $z = 0$ is the same as that from $z = 0$ to $z = .50$. The latter area is found in Table IV to be $.1915$; consequently, the desired area is the sum of these two areas, or $.6853$. Although the two normal curves in Figs. 6 and 7 would look quite different if the same scale were used on both axes, they purposely have been drawn to look alike so that the equivalence of corresponding areas will be apparent.

4. NORMAL APPROXIMATION TO BINOMIAL

Problems related to the binomial distribution are fairly easy to solve provided the number of trials, n, is not large. If n is large, the computations involved in using formula (1) become exceedingly lengthy; consequently, a good simple approximation to the distribution should prove to be very useful. Such an approximation exists in the form of the proper normal distribution. For the purpose of investigating this approximation, consider some numerical examples.

Let $n = 12$ and $p = \frac{1}{3}$ and construct the graph of the corresponding binomial distribution. By the use of formula (1), the values of $P\{x\}$ were computed, correct to three decimals, as

(5)

$$
\begin{array}{lll}
P\{0\} = .008 & P\{4\} = .238 & P\{8\} \;\; = .015 \\
P\{1\} = .046 & P\{5\} = .191 & P\{9\} \;\; = .003 \\
P\{2\} = .127 & P\{6\} = .111 & P\{10\} = .000 \\
P\{3\} = .212 & P\{7\} = .048 & P\{11\} = .000 \\
& & P\{12\} = .000.
\end{array}
$$

Although the graph used earlier for a binomial distribution was a line graph because of the discrete character of the variable x, this distribution will be graphed as a histogram in order to compare it more readily with normal distribution histograms. The graph of the histogram for this distribution is shown in Fig. 8. The height of any rectangle is equal to the probability given by (5) for the corresponding class mark. Since the base length of any rectangle is 1, the area of any rectangle is equal to its height, and therefore these probabilities are also given by the areas of the corresponding rectangles. From the shape of this histogram it appears that it could be fitted fairly well by the proper normal curve.

Since a normal distribution curve is completely determined by its mean and

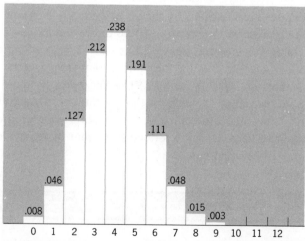

FIGURE 8.　Binomial distribution for $p = \frac{1}{3}$ and $n = 12$.

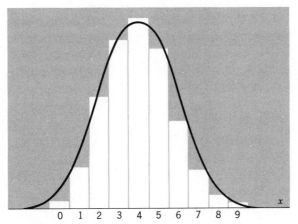

FIGURE 9. Binomial distribution for $p = \frac{1}{3}$ and $n = 12$, with a fitted normal curve.

standard deviation, the natural normal curve to use here is the one with the same mean and standard deviation as the binomial distribution. From the formulas given in (2), it follows that $\mu = 12 \cdot \frac{1}{3} = 4$ and $\sigma = \sqrt{12 \cdot \frac{1}{3} \cdot \frac{2}{3}} = 1.63$. A normal curve with this mean and standard deviation was superimposed on Fig. 8 to give Fig. 9. It appears that the fit is fairly good in spite of the fact that $n = 12$ is a small value of n and advanced theory promises a good fit only for large values of n.

As a test of the accuracy of the normal curve approximation here and as an illustration of how to use normal curve methods for approximating binomial probabilities, consider a few problems related to Fig. 9.

If the probability that a marksman will hit a target is $\frac{1}{3}$ and if he takes 12 shots, what is the probability that he will score at least 6 hits? The exact answer, correct to three decimals, is obtained by adding the values in (5) from $x = 6$ to $x = 12$, which is found to be .177. Geometrically, this answer is the area of that part of the histogram in Fig. 9 lying to the right of $x = 5.5$. Therefore, to approximate this probability by normal curve methods, it is merely necessary to find the area under that part of the fitted normal curve lying to the right of 5.5. Since the fitted normal curve has $\mu = 4$ and $\sigma = 1.63$, it follows that

$$z = \frac{x - \mu}{\sigma} = \frac{5.5 - 4}{1.63} = 0.92.$$

Now, from Table IV, the area to the right of $z = 0.92$ is .179; therefore this is the desired approximation to the probability of getting at least 6 hits. Since the exact answer was just computed to be .177, the normal curve approximation here is certainly good.

To test the accuracy of normal curve methods over a shorter interval, calculate the probability that the marksman will score precisely six hits in twelve shots. From (5) the answer, correct to three decimals, is .111. Since this is equal to the area of the rectangle whose base runs from 5.5 to 6.5, to approximate this answer it is necessary to find the area under the fitted normal curve between $x = 5.5$ and $x = 6.5$. Thus, by calculating the z values and using Table IV, one obtains

$$z_2 = \frac{6.5 - 4}{1.63} = 1.53, \qquad A_2 = .4370$$

$$z_1 = \frac{5.5 - 4}{1.63} = 0.92, \qquad A_1 = .3212.$$

Subtracting these two areas gives .116, which, compared to the exact probability value of .111, is also good. From these two examples it appears that normal curve methods give good approximations even for some situations, such as the one considered here, in which n is not very large.

Suppose, now, that the value of $p = \frac{1}{3}$ is not changed but n is allowed to increase in size. The resulting histogram, like the one in Fig. 9, will move off to the right, spread out, and decrease in height. It is difficult to inspect such a histogram and to observe whether the proper normal curve would fit it well. These undesired changes in the histogram can be prevented by shifting to the corresponding variable in standard units. From (4) this means graphing the histogram for the variable

$$z = \frac{x - \mu}{\sigma}.$$

When formulas (2) are used, the standard variable z assumes the form

(6)
$$z = \frac{x - np}{\sqrt{npq}}.$$

Since the variable z possesses a distribution with mean 0 and standard deviation 1, the histogram for z will behave itself and not go wandering off and flatten out when n becomes large, as is the case of the histogram for x.

Figures 10 and 11 show the histograms for the variable z when $p = \frac{1}{3}$ and $n = 24$ and 48, respectively. They show how rapidly the distribution of z approaches the distribution of a normal variable with mean 0 and standard deviation 1. It can be shown by advanced methods that if p is held fixed and n is allowed to become increasingly large, then the distribution of z will come increasingly close to the

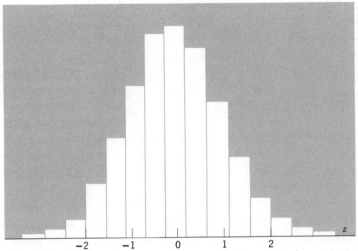

FIGURE 10. Binomial distribution of $(x - np)/$
$\sqrt{npq}$ for $p = \frac{1}{3}$ and $n = 24$.

distribution of a normal variable with mean 0 and standard deviation 1. From a practical point of view, experience has shown that the approximation is fairly good as long as $np > 5$ when $p \leq \frac{1}{2}$ and $nq > 5$ when $p > \frac{1}{2}$.

The fact that the standard form of a binomial variable possesses a distribution approaching that of a standard normal variable implies that the binomial variable

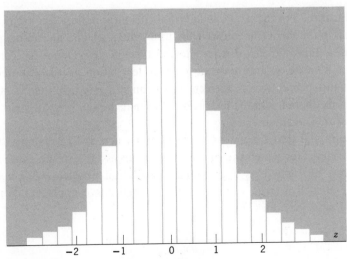

FIGURE 11. Binomial distribution of $(x - np)/$
$\sqrt{npq}$ for $p = \frac{1}{3}$ and $n = 48$.

x possesses a histogram that can be fitted well by the proper normal curve when n is large. The proper normal curve is, of course, the one with mean and standard deviation given by formulas (2).

There are numerous occasions when it is more convenient to work with the proportion of successes, x/n, in n trials than with the actual number of successes, x. If the numerator and denominator in (6) are divided by n, z will assume the form

$$(7) \qquad\qquad z = \frac{\dfrac{x}{n} - p}{\sqrt{\dfrac{pq}{n}}}.$$

The value of z has not changed, only the form in which it is written; consequently z still possesses an approximate normal distribution with zero mean and unit standard deviation when n is large. This implies that the proportion of successes, x/n, possesses a histogram that can be fitted well by the proper normal curve when n is large. The proper normal curve is now the one with mean and standard deviation given by the formulas

$$(8) \qquad\qquad \mu = p$$
$$\sigma = \sqrt{\frac{pq}{n}}.$$

As an illustration of the use of the normal curve approximation to the binomial distribution in the form (7), consider the following problem. Suppose a politician claims that a survey in his district showed that 60 per cent of his constituents agreed with his vote on an important piece of legislation. If it is assumed temporarily that this percentage is correct, and if an impartial sample of 400 voters is taken in his district, what is the probability that the sample will yield less than 50 per cent in agreement?

If it is assumed that taking a sample of 400 voters is like playing a game of chance 400 times for which the probability of success in a single game is .6, this problem can be treated as a binomial distribution problem with $p = .6$ and $n = 400$. For such a large value of n the normal curve approximation will be excellent. Using formula (7),

$$z = \frac{.5 - .6}{\sqrt{\dfrac{(.6)(.4)}{400}}} = -4.08.$$

Now the sample proportion, x/n, will be less than .5 provided that z is less than -4.08. The probability that $z < -4.08$ is by symmetry equal to the probability that $z > 4.08$, which is considered too small to be worth listing in Table IV. Thus, if it should happen that less than 50 per cent of the sample favored the politician, his claim of 60 per cent backing would certainly be discredited.

Objections may be raised, and rightfully so, that getting a sample of 400 voters is not equivalent to playing a game of chance 400 times. There are questions concerning the independence of the trials and the constancy of the probability that must be answered before one can be thoroughly happy with the binomial distribution model for this problem.

As a second illustration of the use of formula (7), the following problem will be solved. Past experience with an examination in freshman English has shown that only 50 per cent of the students pass it. If a new class of 200 freshmen takes the examination, what is the probability that at least 55 per cent will pass it? Here

$$z = \frac{.55 - 50}{\sqrt{\dfrac{(.5)(.5)}{200}}} = 1.41.$$

But $x \geq .55$ if, and only if $z \geq 1.41$; therefore by Table IV the probability of $x \geq .55$ is .08.

As a final illustration, consider the following problem. Suppose that 5 per cent of the individuals who are inoculated against influenza receive rather serious undesirable reactions to the inoculation. Using the normal approximation, calculate the probability that more than 8 per cent of 200 inoculated individuals will receive such reactions.

If formula (7) is applied, one first calculates

$$z = \frac{.08 - .05}{\sqrt{\dfrac{(.05)(.95)}{200}}} = \frac{.03}{\sqrt{.00\ 02\ 37}} = \frac{.03}{.0154} = 1.95.$$

From Table IV it then follows that $P\{z > 1.95\} = .0256$, which is the desired probability that more than 8 per cent will receive reactions.

It should not be assumed from the preceding examples that all binomial distribution problems can be treated satisfactorily by means of the normal approximation even though n may be fairly large. For example, if $n = 80$ and $p = \frac{1}{20}$, calculation of $P\{x\}$ values and a graph of those values will show that the distribution is not sufficiently symmetrical to permit a good fit by a normal curve. Since the mean is 4 here, and x cannot assume negative values, too much of the distribution is concentrated near zero for symmetry. Considerations such as these gave rise to the empirical rule for a good approximation, stated in the paragraph following (6).

▶5.   BINOMIAL DISTRIBUTION DERIVATION

For those who have studied section 7 of Chapter 3, the correctness of formula (1) can be demonstrated as follows.

Consider an experiment of n trials and a particular sequence of outcomes that will produce precisely x successes and $n - x$ failures. One such sequence is the following in which all the successes occurred first, followed by all the failures:

$$\overbrace{SS \cdots S}^{x} \quad \overbrace{FF \cdots F}^{n-x}.$$

Another such sequence is the following one in which a failure occurred first, followed by x consecutive successes, then by the remaining failures. Thus

$$\overbrace{FSS \cdots S}^{x} \quad \overbrace{FF \cdots F}^{n-x-1}.$$

Because of the independence of the trials, the probability of obtaining the first of these two sequences is given by

$$\overbrace{p \cdot p \cdots p}^{x} \cdot \overbrace{q \cdot q \cdots q}^{n-x} = p^x q^{n-x}$$

The probability for the second sequence is given by

$$\overbrace{q \cdot p \cdot p \cdots p}^{x} \cdot \overbrace{q \cdot q \cdots q}^{n-x-1} = p^x q^{n-x}$$

Thus the probability for the two sequences is the same and will be the same for every sequence that satisfies the condition of having x successes and $n - x$ failures.

The number of ways in which the desired event can occur is equal to the number of different sequences that can be written down of the type just displayed, those containing x letters S and $n - x$ letters F. But this is equal to the number of ways of choosing x positions out of n available positions in which to place the letter S. The remaining $n - x$ positions will automatically be assigned the letter

F. The n positions may be numbered and treated like n numbered cards. The problem then is to choose x of those cards. Since only the numbers on the cards are of interest and not the order in which they are drawn, this is a combination problem. From the derivation in section 7 of Chapter 3, the number of ways of choosing x cards from n distinct cards is given by the combination formula (13) of Chapter 3, namely,

$$\binom{n}{x} = \frac{n!}{x!(n-x)!}.$$

This, therefore, is the number of sequences that produce exactly x successes.

Since each of these sequences represents one of the mutually exclusive ways in which the desired event can occur, and each such sequence has the same probability of occurring, namely $p^x q^{n-x}$, it follows that the desired probability is obtained by adding this probability as many times as there are sequences. But the number of such sequences was just found to be $\binom{n}{x}$; therefore $P\{x\}$ is obtained by multiplying $p^x q^{n-x}$ by $\binom{n}{x}$, which verifies formula (1).

6.   ADDITIONAL ILLUSTRATIONS

1.   Suppose 60 per cent of the students at a large university work part time. Let x denote the number of such students in a sample of 8 students chosen by chance from the registration files. Assuming that the size of the university is so large that $p = .6$ may be assumed to be constant when a sample of only 8 students is taken, (a) 1 find an expression for $P\{x\}$ which gives the distribution of x. (b) 1 Use the result in (a) to calculate $P\{x \geq 6\}$. (c) 2 Use formulas (2) to calculate the mean and standard deviation of x. (d) 4 Use the normal approximation to approximate the value found in (b). The solutions follow.

(a) The sample of 8 may be treated as eight independent trials of an experiment for which $p = .6$; hence x is a binomial variable with distribution given by

$$P\{x\} = \frac{8!}{x!(8-x)!}(.6)^x(.4)^{8-x}$$

(b) $P\{x \geq 6\} = P\{6\} + P\{7\} + P\{8\}$

$$= \frac{8!}{6!2!}(.6)^6(.4)^2 + \frac{8!}{7!1!}(.6)^7(.4)^1 + \frac{8!}{8!0!}(.6)^8(.4)^0$$

$$= (.6)^6[28(.4)^2 + 8(.6)(.4) + (.6)^2]$$

$$= .315.$$

(c) $\mu = 8(.6) = 4.8$, $\sigma = \sqrt{8(.6)(.4)} = 1.39$

(d) $z = \dfrac{5.5 - 4.8}{1.39} = .50$, $P\{z > .50\} = .31$.

2.   A box contains the following nine cards: the three, four, and five of spades, the three and four of clubs, the three and four of hearts, and the four and five of diamonds. If a card is to be drawn from the box and the experiment is to be repeated 10 times with the drawn card always being replaced and if x denotes the number of black cards that will be obtained, (a) 1 find an expression for $P\{x\}$ which gives the distribution of x. (b) 1 Use the result in (a) to calculate $P\{4 \leq x \leq 5\}$. (c) 4 Use the normal approximation to approximate the value found in (b). (d) 4 Calculate the approximate probability, using the normal approximation, that the proportion of black cards in 100 trials of the experiment will be less than .6.

(a) $n = 10$, $p = \dfrac{5}{9}$

$$P\{x\} = \frac{10!}{x!(10-x)!}\left(\frac{5}{9}\right)^{x}\left(\frac{4}{9}\right)^{10-x}$$

(b) $P\{4 \leq x \leq 5\} = P\{4\} + P\{5\}$

$$= \frac{10!}{4!6!}\left(\frac{5}{9}\right)^{4}\left(\frac{4}{9}\right)^{6} + \frac{10!}{5!5!}\left(\frac{5}{9}\right)^{5}\left(\frac{4}{9}\right)^{5}$$

$$= .39.$$

(c) $\mu = np = 10 \cdot \dfrac{5}{9} = \dfrac{50}{9} = 5.56$

$\sigma = \sqrt{npq} = \sqrt{10 \cdot \dfrac{5}{9} \cdot \dfrac{4}{9}} = \dfrac{10}{9}\sqrt{2} = 1.57$

$z_1 = \dfrac{5.50 - 5.56}{1.57} = -\dfrac{.06}{1.57} = -.04$

$z_2 = \dfrac{3.50 - 5.56}{1.57} = -\dfrac{2.06}{1.57} = -1.31$

$A_1 = .0160$, $A_2 = .4049$, $A_2 - A_1 = .3889$; hence

$$P\{4 \leq x \leq 5\} \doteq .3889, \text{ or } .39.$$

(d) $z = \dfrac{\dfrac{x}{n} - p}{\sqrt{\dfrac{pq}{n}}} = \dfrac{.60 - .556}{\sqrt{\dfrac{\frac{5}{9} \cdot \frac{4}{9}}{100}}} = \dfrac{.044}{.050} = .88$

$$A = .31; \text{ hence } P\left\{\frac{x}{10} < .6\right\} = .50 + .31 = .81.$$

3. Let x represent the weight in pounds of a king salmon caught at the mouth of a certain river and assume that x possesses a normal distribution with mean 30 and standard deviation 6. Calculate the probability that if a fisherman catches a salmon its weight will (a) 3 be at least 41 pounds, (b) 3 be between 20 and 40 pounds, inclusive.

(a) $z = \dfrac{40.5 - 30}{6} = 1.75$, $P\{x \geq 41\} = P\{z \geq 1.75\} = .04$

(b) $z_1 = \dfrac{19.5 - 30}{6} = -1.75$, $A_2 = \dfrac{40.5 - 30}{6} = 1.75$,

$P\{20 \leq x \leq 40\} = P\{-1.75 \leq z \leq 1.75\} = 2P\{0 \leq z \leq 1.75\} = .92$

EXERCISES

Section 1

1. Verify the values given in the text in Table 3 by means of formula (1).

2. Use the method of enumeration of possible outcomes with their corresponding probabilities to derive the binomial distribution for the number of heads obtained in tossing a coin 4 times. Check your results by means of formula (1).

3. Work problem 2 for the number of aces obtained in rolling a die 4 times.

4. A coin is tossed 5 times. Using formula (1), calculate the values of $P\{x\}$, where x denotes the number of heads, and graph $P\{x\}$ as a line chart.

5. If the probability that you will win a hand of bridge is $\frac{1}{4}$ and you play 6 hands, calculate the values of $P\{x\}$, where x denotes the number of wins, by means of formula (1).

6. If 20 per cent of fuses are defective, and a box of ten fuses is purchased, what is the probability that at least seven of those fuses will be good?

7. In problem 6 how many fuses would you need to buy so that you would have a probability of at least .9 of obtaining at least seven good fuses?

8. Would you expect the binomial distribution to be applicable to a calculation of the probability that the stock market will rise at least 20 of the days during the next month if you have a record for the last 5 years of the percentage of days that it did rise? Explain.

9. Explain why it would not be strictly correct to apply the binomial distribution to a calculation of the probability that it will rain at least 10 days during next January if each day in January is treated as a trial of an event and one has a record of the percentage of rainy days in January.

Section 2

10. For the binomial variable for which $n = 8$ and $p = \frac{1}{4}$, calculate the mean and standard deviation and verify your results by means of formulas (2).

11. For problem 5, calculate the mean and standard deviation for the variable x and verify your results by means of formulas (2).

►**12.** Derive the formula $\mu = np$ for the general binomial distribution $P\{x\}$ given by (1) by writing out the terms in $\Sigma_{x=0}^{n} xP\{x\}$, then factoring out the common factor np, then calculating $\Sigma_{x=0}^{n-1} Q(x)$ where $Q(x)$ is the same as $P\{x\}$ for $n - 1$ trials, and finally recognizing that this last sum, which has the value 1, is the other factor in $\Sigma xP\{x\}$. The formula seems obvious but its algebraic derivation is not.

►**13.** Consult a more advanced statistics text to observe the algebraic technique used to derive the formula $\sigma = \sqrt{npq}$.

Section 3

14. Given that x is normally distributed with mean 12 and standard deviation 2, use Table IV to calculate the probability that (a) $x > 14$, (b) $x > 11$, (c) $x < 10$, (d) $x < 10.5$, (e) $10 < x < 13$.

15. Assuming that stature (x) of college males is normally distributed with mean 69 inches and standard deviation 3 inches, use Table IV to calculate the probability that (a) $x < 66$ inches, (b) 65 inches $< x < 71$ inches.

16. Suppose your score on an examination in standard units (z) is .9 and scores are assumed to be normally distributed. What percentage of the students would be expected to score higher than you?

17. A high school gym teacher announces that he grades individual athletic events by achievement relative to all his classes. If he gives 20 per cent A's and if experience has shown that the mean is 4 feet 9 inches and the standard deviation is 4 inches for the high jump, how high should a student plan to jump if he expects to get an A?

18. Suppose $n = 20$ and $p = \frac{1}{10}$. Calculate $P\{0\}$, and on the basis of its value argue that a good normal curve fit to the entire distribution would not be expected here.

19. Assuming that the I.Q. scores of the students at a college are normally distributed with a mean of 112 and a standard deviation of 8, calculate the percentage of students who will have an I.Q. score (a) higher than 130, (b) lower than 100, (c) between 105 and 125, inclusive.

20. Experience with a placement examination in basic English indicates that examination scores are approximately normally distributed with mean 120 and standard deviation 20. If a score of 100 is required to pass the examination, what percentage of the students can be expected to fail the examination?

Section 4

21. A coin is tossed 8 times. Find the probability, both exactly by means of formula

. ||

(1) and approximately by means of the normal curve approximation, of getting (a) 6 heads, (b) at least 6 heads. *. 14*

22. If the probability of your winning at pinochle is .4, find the probability, both exactly by means of formula (1) and approximately by means of the normal curve approximation, of winning 4 or more of 7 games played.

23. If 30 per cent of students have defective vision, what is the probability that at least half of the members of a class of 20 students will possess defective vision? Use the normal curve approximation.

24. If 10 per cent of television picture tubes burn out before their guarantee has expired, (a) what is the probability that a merchant who has sold 100 such tubes will be forced to replace at least 15 of them? (b) What is the probability that he will replace at least 5 and not more than 15 tubes? Use the normal curve approximation.

.04

25. If 20 per cent of the drivers in a certain city have at least 1 accident during a year's driving, what is the probability that the percentage for 200 customers of an insurance company in that city will exceed 25 per cent during the next year? Use the normal curve approximation.   *.04   .03*

26. If 10 per cent of the cotter pins being manufactured are defective, what is the approximate probability that the percentage of defectives in a box of 200 will exceed 16 per cent?

27. For problem 18, calculate $P\{x \le 0\}$ by means of the normal curve approximation and compare with the result in problem 18.

28. Give an example of a binomial distribution for which $n > 100$ and p is not small but for which there will not be a good normal approximation.

Section 6

29. A box contains four black cards numbered 1, 2, 3, and 4, three red cards numbered 2, 3, and 4, and three white cards numbered 3, 4, and 5. Work parts (a) through (d) of the illustrative review exercise 2 in section 6. (e) If the experiment in (a) is repeated eight times, with the drawn card always replaced, and x denotes the number of black cards obtained, find an expression for $P\{x\}$. (f) Use the result in (e) to calculate $P\{3 \le x \le 5\}$. (g) Use the normal approximation to solve (f). (h) Calculate the probability, using the normal approximation, that the proportion of black cards will be greater than $\frac{1}{2}$ if the experiment is conducted 50 times.

23. $\dfrac{9.5 - .6}{2.04}$ 25.) $\dfrac{50.5 - 40}{5.66}$

mean μ
s. dev. σ

$z = \dfrac{x - \mu}{\sigma}$

CHAPTER | 6

Sampling

1. INTRODUCTION

The preceding chapters have been concerned with describing frequency distributions obtained from sampling some population and with constructing probability distributions that may serve to represent the population. It has been assumed that the population distribution can be approximated with high accuracy if the sample is sufficiently large, and therefore that the sample frequency distribution represents the population being sampled satisfactorily. For example, if a manufacturer of shirts takes a large sample of neck sizes of adult males, he assumes that the distribution of those sizes for the sample will be very close to the distribution of such sizes for all adult males in the population. If this were not true, the sample could not serve as a satisfactory basis for making statistical inferences about the population being sampled.

The problem of how to extract samples so that valid inferences concerning a population can be made was discussed very briefly in Chapter 2. There it was stated that a method of sampling called random sampling was satisfactory from this point of view. The properties of random sampling that enable it to yield valid inferences will now be discussed and efficient methods for drawing random samples will be described. After these preliminaries have been disposed of, the general problem of determining how the accuracy of a sample depends on the size of the sample will be studied. This study will involve both theoretical results and several sampling experiments. The purpose of the experiments is to demonstrate sampling variability and to make the theoretical results more plausible.

2. RANDOM SAMPLING

In Chapter 2, random sampling was defined as a sampling procedure in which every member of the population has the same chance of being selected. In terms of probability this implies that the probability of any particular member being selected is $1/N$, where N denotes the number of individuals in the population. More generally, if the sample is to contain r individuals, in which case the sample is said to be of size r, the sampling is defined to be random if every combination of r individuals in the population has the same chance of being selected.

Although random sampling was advocated earlier on the grounds that it is a method of sampling for which one can expect the sample distribution to represent the population distribution in miniature, there are more important reasons for advocating it. The most important of these is that random sampling leads to probability models for distributions. Since the conclusions to be drawn about populations by means of samples are to be based upon probabilities, samples must be selected in such a manner that the rules of probability can be applied to them. For the purpose of seeing why this is so, return to the problem discussed in the preceding chapter concerning a sample of 400 voters. There it was assumed that the sample of 400 voters could be treated as 400 independent trials of an experiment for which the probability of success in a single trial is $p = .6$. This assumption permitted the problem to be treated as a binomial distribution problem and therefore permitted the calculation of probabilities of various possible outcomes by means of the binomial distribution and its normal approximation.

The two features that are necessary to enable the binomial distribution to be applied to practical problems such as this are the independence of the trials and the constant probability for all trials. For a finite population of N individuals, the probability that an individual selected at random will favor a certain candidate will not necessarily remain constant as successive samples are taken; however, if N is very large compared to the size of the sample, the proportion of individuals favoring the candidate will change very little as the sampling progresses. Therefore, if the samples are chosen so that at each stage of the sampling each member remaining in the population has the same chance of being drawn, the conditions for a binomial model will be satisfactorily realized. Thus, random sampling will permit the use of a binomial distribution model for such problems provided N is large. If N is not sufficiently large to justify this usage, a more sophisticated probability model can be employed; however, it will not be discussed here.

If some other type of sampling had been employed to obtain the sample of 400 voters, there would be no assurance that a binomial distribution model would be valid for calculating probabilities. For example, if a sampler started down a residential street and sampled the first 400 houses on that street he might well encounter a group of individuals whose similarity of economic status would tend to make the proportion of them favoring the candidate considerably higher, or lower, than the proportion for the entire city. One cannot, in general, make valid probability statements about the outcomes of other types of sampling methods. It is for this reason that statisticians insist that samples be randomly selected.

There are many examples in public life of fiascos that have occurred because conclusions were based on nonrandom samples. For example, in 1936 the *Literary Digest* conducted a poll by mail for the purpose of predicting the forthcoming presidential election. It sent out 10,000,000 ballots and on the basis of more than 2,000,000 returns predicted that Landon would be elected. Actually, Roosevelt received approximately 60 per cent of the votes cast in that election. Newspapers regularly report the opinions and predictions of politicians as though their statements possessed much merit, even when the politician declares that his opinions are based upon "sounding out" his constituents. Businessmen frequently make incorrect decisions in such fields as marketing when their decisions are based on faulty information obtained from poor samples.

Random samples can be obtained by employing a game-of-chance technique. For example, if there were 10,000 voters in a city and a random sample of size 400 were to be taken, it would suffice to write the name of each voter on a slip of paper, mix the slips well in a large container, and then draw 400 slips from the container, which is thoroughly mixed after each individual drawing.

A less cumbersome and cheaper method of selecting a random sample is to employ a table of random numbers. Such a table could be constructed by writing the digits 0 to 9 on ten slips of paper, mixing them thoroughly between each drawing, replacing the slip drawn each time, and recording the digits so obtained. The resulting sequence of random digits could then be used to construct random numbers of, say, five digits each. The table of random numbers given in Table II of the appendix could have been obtained in this fashion, but it was obtained by a more refined and foolproof method. Suppose now that the 10,000 voters are listed in a register of voters and each voter is associated with a number from 0000 to 9999. Then to obtain a sample of 400 names it is merely necessary to select 400 sets of four-digit numbers from Table II. Since these numbers occur in groups of five, one would select only the first four digits of a group to yield a name. There are fifty rows of numbers in each column of such numbers; therefore, eight such columns would suffice. From the manner in which random numbers are formed, it follows that every four-digit number has the same probability of being formed at any specified place in the table; therefore, one can just as well choose the numbers systematically by reading down a column as by jumping around in the table. It could happen, of course, that one or more individuals will be selected more than once when random sampling numbers are used because these numbers were formed independently, and hence a particular four-digit number may occur more than once in the table. A few extra random numbers must then be chosen to fill out the sample. If there had been only 7500 registered voters in the city, one would discard any four-digit number obtained from the table that is larger than 7499 because there would be no voters associated with those numbers.

3. DISTRIBUTION OF $\bar{x}$ WHEN SAMPLING A NORMAL POPULATION

The purpose of taking a random sample from a population is, of course, to obtain information about the population distribution. That information is usually expressed in the form of an empirical frequency distribution with its accompanying histogram, or in the form of arithmetical descriptive quantities such as $\bar{x}$ and s^2. Thus, an empirical distribution is a sample estimate of the probability distribution representing the population. Similarly, $\bar{x}$ and s^2 are sample estimates of the parameters μ and σ^2 of the population probability distribution.

Since most of the statistical inferences that will be made about populations by means of samples will be concerned with such quantities as the mean and variance, it is time to begin the study of the reliability of the sample mean $\bar{x}$ and sample variance s^2 as estimates of the corresponding population parameter values μ and σ^2. This chapter will be concerned principally with studying properties of the sample mean because it is the more important of the two and is considerably easier to treat than the sample variance.

The problem of determining the accuracy of a sample estimate of a population parameter can be broken down into two parts. One is concerned with determining whether the estimate is biased. For example, if an estimate of the average height of adult males were obtained from a random sample of college students the value of $\bar{x}$ would tend to be larger than μ because college students tend to be larger than the rest of the older adult population. An estimate that tends to overestimate, or underestimate, a parameter value is said to be a biased estimate. A more precise definition will be given shortly. The second part of the problem is concerned with determining the precision of an estimate. If an estimate is unbiased, then precision and accuracy mean the same thing; however an estimate may be very precise and heavily biased, with the result that is not an accurate estimate. Precision refers to the property of yielding values under repeated sampling which are very close together.

For the purpose of discussing the property of an estimate being unbiased, consider the experiment of taking a random sample of size 10 from some population whose probability distribution has the mean μ. Let $\bar{x}$ denote the mean of the sample. If this experiment were repeated a large number of times, a large number of values of $\bar{x}$ would be available for obtaining an empirical distribution of the variable $\bar{x}$. The mean of this empirical distribution would be expected to approach some fixed value as the experiment is repeated indefinitely. If the value being approached turns out to be μ, then $\bar{x}$ is said to be an unbiased estimate of μ. Stated somewhat differently, $\bar{x}$ is an unbiased estimate of μ provided that the mean of its probability distribution is μ. Since the expected value E of a random variable is the mean of its probability distribution, it follows that $\bar{x}$ will be an *unbiased estimate* of μ if, and only if, its expected value is equal to μ. This definition applies to other parameters and their corresponding estimates as well.

Now it can be shown by using the three properties of E given in Chapter 4 that

$$E[\bar{x}] = \mu \quad \text{and} \quad E[s^2] = \sigma^2.$$

Thus, $\bar{x}$ is an unbiased estimate of μ, and s^2 is an unbiased estimate of σ^2. This latter fact is the justification for dividing $\sum_{i=1}^{n} (x_i - \bar{x})^2$ by $n - 1$ rather than by n in the definition of s^2 given by formula (7), Chapter 2. If n had been used in the denominator, s^2 would have been a biased estimate of σ^2 and would have produced an estimate that is smaller than σ^2 on the average.

In view of the fact that $\bar{x}$ and s^2 are unbiased estimates of μ and σ^2, the study of their accuracy is reduced to studying how those estimates vary about μ and σ^2, respectively, under repeated sampling experiments. Since $\bar{x}$ is more useful than s^2 in statistical inference problems and since its sampling distribution is easier to study than that of s^2, only the distribution of $\bar{x}$ will be studied in this chapter.

There are two basic mathematical theorems on the accuracy of $\bar{x}$. Rather than merely state them, two sampling experiments will be carried out to observe how $\bar{x}$ varies in repeated sampling and to verify that the sampling results agree with those predicted by the theorems.

In the first experiment, samples of size 4 will be taken from the population whose probability distribution is given by Fig. 1. This distribution is a discrete approximation to the standard normal distribution. It was obtained by using Table IV to calculate the percentage of area under the standard normal curve over unit intervals with the middle interval centered at the origin. Incidentally, the percentages obtained from Fig. 1, when the two end interval percentages are combined with their neighboring intervals on each side, give the percentages that are often used by instructors who "grade on the curve" to determine the percentage of letter grades A, B, C, D, and F to assign.

Samples of size 4 were taken from this discrete distribution by means of the random numbers found in Table II. First, a tabulating form of the type shown in Table 1 was constructed. By this procedure all pairs of random digits were divided into seven groups according to the proportions shown in Fig. 1 and associated with the class marks of Fig. 1. For example, $x = -3$ is assigned to the pair 00, which is 1 per cent of all such pairs, and the value $x = -2$ is assigned to the pairs from 01 to 06 inclusive, which include 6 per cent of all random-number pairs. Four such pairs of random numbers are read from Table II and recorded in the proper class interval to form one experiment. The results of the first three such experiments are shown in Table 1. This experiment was repeated 100 times. Next, the mean for each experiment was calculated and recorded in the last row of Table 1, labeled $\bar{x}$.

The values of $\bar{x}$ were next tabulated to yield the frequencies shown in Table 2

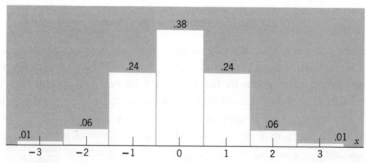

FIGURE 1. An approximation to the standard normal distribution.

TABLE 1

x	Random numbers	1	2	3	...
-3	00				
-2	01–06		/	/	
-1	07–30	/			
0	31–68	//	/	///	
1	69–92	/	/		
2	93–98		/		
3	99				
$\bar{x}$		0	$\frac{1}{4}$	$-\frac{1}{2}$	...

in which the third row gives the percentages in decimal fraction form of the corresponding absolute frequencies.

Finally, this frequency table was graphed as a histogram. Since total areas must be equal to 1 in comparing probability distributions, and since the class interval in this table is $\frac{1}{4}$, it is necessary to draw rectangles that are four times as high as the $f/100$ values. The resulting histogram, with area 1, is shown in Fig. 2.

On comparing Figs. 1 and 2, it is apparent that sample means based on four measurements each do not vary as much as do individual sample values. This is certainly to be expected because, for example, a large value of $\bar{x}$ would require four large values of x, and the probability of getting four large values is much smaller than the probability of getting one large value. The standard deviation of the $\bar{x}$ distribution is obviously considerably smaller than that for the x distribution. Furthermore, it is apparent that the $\bar{x}$ distribution possesses a mean that is close to 0, which is the mean of the x distribution. Finally, it appears that the distribution of $\bar{x}$, except for the difference in spread, possesses a distribution of the same approximate normal type as the x distribution.

TABLE 2

$\bar{x}$	$-\frac{5}{4}$	$-\frac{4}{4}$	$-\frac{3}{4}$	$-\frac{2}{4}$	$-\frac{1}{4}$	0	$\frac{1}{4}$	$\frac{2}{4}$	$\frac{3}{4}$	$\frac{4}{4}$	$\frac{5}{4}$
f	1	4	8	11	15	19	18	12	5	4	3
$\dfrac{f}{100}$	.01	.04	.08	.11	.15	.19	.18	.12	.05	.04	.03

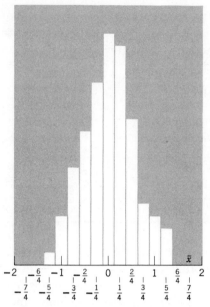

FIGURE 2. Distribution of $\bar{x}$ for samples of size 4 from the distribution of Figure 1.

From Fig. 1 it is clear that $\mu = 0$. Calculations based on formula (2), Chapter 4, and the probabilities given by Fig. 1 will show that $\sigma = 1.07$. If the distribution of Fig. 1 had been that of a continuous standard normal variable instead of a discrete approximation to it the value of σ would have been exactly 1.

Calculations of the mean and standard deviation for the empirical $\bar{x}$ distribution given by Table 2 were made by means of formulas (4) and (8) of Chapter 2 and produced the values .015 and .545, respectively. Thus, the mean of the $\bar{x}$ distribution is very close to the mean, $\mu = 0$, of the x distribution whereas the standard deviation of the $\bar{x}$ distribution is approximately one-half as large as the standard deviation, $\sigma = 1.07$, of the x distribution.

If this sampling experiment had been carried out, say, 500 times rather than just 50, irregularities such as those in Fig. 2 would disappear and it would be found that the properties of the $\bar{x}$ distribution just discussed would become increasingly apparent. Thus, it would be found that the histogram could be fitted very well with a normal curve, that the mean of the $\bar{x}$ distribution would be very close to 0, and that the standard deviation of the $\bar{x}$ distribution would have a value very close to one-half the value of the standard deviation of the x distribution. Although the samples here were taken from an approximate normal distribution as given by Fig. 1 rather than from an exact normal distribution, similar results would be obtained if the approximation were made increasingly good by choosing a very small class interval.

Fortunately, it is not necessary to carry out such repeated sampling experiments to arrive at the theoretical distribution for $\bar{x}$. By using the rules of probability and advanced mathematical methods, it is possible to derive the equation of the curve representing the distribution of $\bar{x}$ when the sampling is from the exact normal distribution rather than an approximation. This corresponds to what was done in Chapter 4 to arrive at the theoretical distribution for binomial x without performing any sampling experiments. It turns out that $\bar{x}$ will possess a normal distribution if x does, with the same mean as x but with a standard deviation that is $1/\sqrt{n}$ times the standard deviation of x. These mathematical results are expressed in the form of a theorem:

(1)   ***Theorem 1.***   *If x possesses a normal distribution with mean μ and standard deviation σ, then the sample mean $\bar{x}$, based on a random sample of size n, will also possess a normal distribution with mean μ and standard deviation $\sigma/\sqrt{n}$.*

The distribution of $\bar{x}$ given by this theorem is often called the sampling distribution of $\bar{x}$ because of its connection with repeated sampling experiments, even though it is derived by purely mathematical methods.

The results of the sampling experiment just completed appear to be in agreement with this theorem. The histogram of Fig. 2 looks like the type of histogram that one gets from samples from a normal population, and since $\sigma \doteq 1$ and $n = 4$ here, its mean and standard deviation are in agreement with the theoretical values of 0 and $\sigma/\sqrt{4} \doteq \frac{1}{2}$ given by the theorem. From this theorem one can draw the conclusion that the means of samples of size 4 from a normal population possess only one-half the variability about the mean of the population that the individual measurements do.

As an illustration of how this theorem can be used to determine the accuracy of $\bar{x}$ as an estimate of μ when x possesses a normal distribution, consider the following problem. Let x represent the height of an individual selected at random from a population of adult males. Assume that x possesses a normal distribution with mean $\mu = 68$ inches and standard deviation $\sigma = 3$ inches. The geometry of this distribution is shown in Fig. 3. Data on male stature show that these assumptions are quite realistic. The problem to be solved is the following one: if a random sample of size $n = 25$ is taken from this population, what is the probability that the sample mean $\bar{x}$ will differ from the population mean by less than one inch?

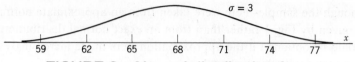

FIGURE 3.   Normal distribution for x.

Since $\mu = 68$, $\sigma = 3$, and $n = 25$, it follows from Theorem 1 that $\bar{x}$ will possess a normal distribution with mean 68 and standard deviation given by

$$\sigma_{\bar{x}} = \frac{\sigma}{\sqrt{n}} = \frac{3}{\sqrt{25}} = .6.$$

The geometry of this distribution is shown in Fig. 4. Now the error of estimate will be less than one inch if $\bar{x}$ falls inside the interval $(67, 69)$. The values of z corresponding to $\bar{x} = 67$ and $\bar{x} = 69$ are given by

$$z_1 = \frac{67 - 68}{.6} = -1.67, \text{ and } z_2 = \frac{69 - 68}{.6} = 1.67.$$

Since, from Table IV, the probability that z will lie between -1.67 and 1.67 is equal to .90, this value is also the probability that x will be in error by at most one inch. Thus, by taking a random sample of size 25 and using the resulting value of $\bar{x}$ to estimate the mean of the entire population, one can be quite certain that his estimate will not differ from the population mean by more than one inch.

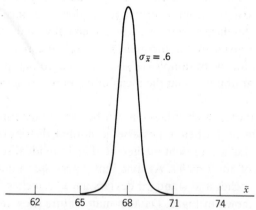

FIGURE 4.   Normal distribution for $\bar{x}$ when $n = 25$.

4.   DISTRIBUTION OF $\bar{x}$ WHEN SAMPLING A NON-NORMAL POPULATION

Suppose now that the variable x does not possess a normal distribution. What then can be said about the distribution of $\bar{x}$? A number of statisticians have conducted sampling experiments with different kinds of nonnormal distributions for x to see what effect the nonnormality would have on the distribution for $\bar{x}$. The surprising result has always been that if n is larger than about 25 the distribution of $\bar{x}$ will appear to be normal in spite of the population distribution chosen for x. This remarkable property of $\bar{x}$ is of much practical importance because a large share of practical problems involve samples sufficiently large to permit one to assume that $\bar{x}$ is normally distributed and thus permit the use of familiar normal curve methods to solve problems related to means without being concerned about the nature of the population distribution.

A well known mathematical theorem, known as a "central limit theorem," essentially states that under very mild assumptions the distribution of $\bar{x}$ will approach a normal distribution as the sample size, n, increases. This theorem can be expressed in the following manner:

(2)   **Theorem 2.**   *If x possesses a distribution with mean μ and standard deviation σ, then the sample mean $\bar{x}$, based on a sample of size n, will have a distribution that approaches the distribution of a normal variable with mean μ and standard deviation $\sigma/\sqrt{n}$ as n becomes infinite.*

For the purpose of demonstrating that $\bar{x}$ will possess an approximate normal distribution with mean μ and standard deviation $\sigma/\sqrt{n}$, even though x is not normally distributed provided that n is sufficiently large, the following sampling experiment was conducted. Random samples of size 10 were chosen from the population represented by the histogram in Fig. 5. The variable x here is a discrete variable that can assume only the integer values 1 to 6, with the corresponding probabilities indicated on the histogram. Theorem 2 does not require that x be a continuous variable; it may be either continuous or discrete. A histogram is used in place of a line chart merely to display the lack of normality better.

Two-digit random numbers from Table II were divided into six groups, corresponding to the six possible values of x. The first 25 per cent of those numbers, namely all those from 00 to 24, were assigned the x value 1. The next 25 per cent, those from 25 to 49, were assigned the x value 2, the next 20 per cent, those from 50 to 69, the x value 3, etc. After these assignments had been made, a column

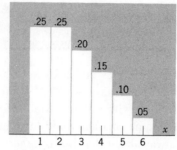

FIGURE 5. Population distribution for sampling experiment.

of two-digit random numbers was selected. The first 10 numbers in that column yielded the first sample of size $n = 10$ from the x population. The second set of 10 numbers in that column yielded the second sample of 10, etc. This was continued, with as many columns as needed, until 100 samples had been drawn.

In the next step of the experiment the values of $\bar{x}$ for those 100 samples were calculated. After the $\bar{x}$ values had been calculated, they were classified in a frequency table in the manner discussed in Chapter 2. The results of this classification are shown in Table 3 and Fig. 6. There was no attempt made to keep the area of the histogram equal to 1 because only the shape of the histogram is of interest here. The histogram of Fig. 6 certainly has the appearance of one that might have been obtained from sampling a normal population. Thus the claim in (2) that $\bar{x}$ should have an approximate normal distribution seems to have been fulfilled here.

TABLE 3

f	1	0	2	7	13	16	19	16	14	9	1	2
$\bar{x}$	1.55	1.75	1.95	2.15	2.35	2.55	2.75	2.95	3.15	3.35	3.55	3.75

Finally, the values of the mean and the standard deviation for the data of Table 3 were calculated by the methods explained in Chapter 2, with the following results:

(3)
$$\text{mean of } \bar{x} \text{ distribution} = 2.77$$
$$\text{standard deviation of } \bar{x} \text{ distribution} = .41.$$

These values need to be compared with the values expected from Theorem 2.

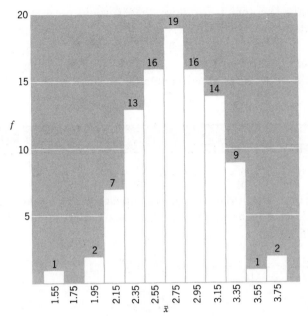

FIGURE 6. Histogram for 100 sample means.

In order to make this comparison, it is necessary to know the values of μ and σ for the x population being sampled. Calculations by means of formulas (1) and (2), Chapter 4, for the probability distribution given by Fig. 5 yielded the values $\mu = 2.75$ and $\sigma = 1.48$, correct to two decimal places. Since $n = 10$ in this experiment, Theorem 2 states that the theoretical normal distribution of $\bar{x}$ will therefore have as mean and standard deviation the quantities

(4)
$$\mu_{\bar{x}} = \mu = 2.75$$

$$\sigma_{\bar{x}} = \frac{\sigma}{\sqrt{n}} = \frac{1.48}{\sqrt{10}} = .47$$

The values obtained from the experiment, given in (3), appear to agree reasonably well with the theoretical values given in (4). The theoretical values given in (4) are those that should be approached by the values in (3) if the sampling experiment were continued indefinitely instead of stopping after 100 such experiments.

Since $n = 10$ is a small sample size and since the population distribution for x is far removed from being normal, one could hardly have expected the distribution of $\bar{x}$ to fit the theory in (2) too well, and yet it appears to do so very well.

Since $n = 25$ is sufficiently large to justify the use of Theorem 2 on the problem used to illustrate Theorem 1, it follows that it would not have been necessary to assume that stature of adult males is normally distributed in order to solve that problem.

The preceding theorems will be used extensively in the following chapters for making statistical inferences about population means.

5.   ADDITIONAL ILLUSTRATIONS

1.   Assume that scores, x, on a placement test are normally distributed with mean 160 and standard deviation 20. If a sample of size 16 is taken and the value of x computed, what is the probability that (a) 3 $\bar{x}$ will exceed 165, (b) 3 $\bar{x}$ will be less than 150?

(a) $\sigma_{\bar{x}} = \dfrac{\sigma}{\sqrt{n}} = \dfrac{20}{\sqrt{16}} = 5, \quad z = \dfrac{165 - 160}{5} = 1,$

$$P\{\bar{x} > 165\} = P\{z > 1\} = .16.$$

(b) $z = \dfrac{150 - 160}{5} = -2,$

$$P\{\bar{x} < 150\} = P\{z < -2\} = .023.$$

2.   Let x represent the grade point average of a randomly selected student from a certain university. It is known that the distribution of x has a mean of 2.5 and a standard deviation of .4. If a sample of 36 students is taken and the value of $\bar{x}$ calculated, what is the probability that $\bar{x}$ will (a) 4 be less than 2.4, (b) 4 be in the interval (2.4, 2.7).

(a) Although grade point averages are not normally distributed, a sample of size 36 justifies the use of Theorem 2; hence

$$\sigma_{\bar{x}} = \frac{.4}{\sqrt{36}} = .067, \, z = \frac{2.4 - 2.5}{.067} = -1.50$$
$$P\{\bar{x} < 2.4\} = P\{z < -1.50\} = .07.$$

(b) $z_1 = \dfrac{2.4 - 2.5}{.067} = -1.50, \, z_2 = \dfrac{2.7 - 2.5}{.067} = 3.00,$

$P\{2.4 < \bar{x} < 2.7\} = P\{-1.50 < z < 3.00\} = P\{0 < z < 1.50\} +$
$$P\{0 < z < 3.00\} = .4332 + .4987 = .93.$$

EXERCISES

Section 2

1. Suggest how to take a random sample of 100 students from the students at a university.

2. Give reasons why taking every tenth name from the names under the letter A in a telephone book might or might not be considered a satisfactory random-sampling scheme for studying the income distribution of adults in a city.

3. Airlines often leave questionnaires in the seat pockets of their planes to obtain information from their customers regarding their services. Criticize this method of obtaining information.

4. During a prolonged debate on an important bill in the United States Senate, Senator A received 300 letters commending him on his stand and 100 letters reprimanding him for his stand. Senator A considered these letters as a fair indication of public sentiment on this bill. Comment on this.

5. A business firm sent out questionnaires to a random sample of 1000 housewives in a certain city concerning their views about paper napkins. Of these, 400 replied. Would these 400 replies be satisfactory for judging the general views of housewives on napkins?

6. How could you use random numbers to take samples of wheat in a wheat field if the wheat field is a square, each side of which is 1000 feet long, and if each sample is taken by choosing a random point in the square and harvesting the wheat inside a hoop 5 feet in diameter whose center is at the random point?

7. An agency wishes to take a sample of 200 adults in a certain residential section of a city. It proposes to do so by taking a random sample of 200 households obtained from a listing of all households in that district and then selecting at random 1 adult from each such household. Why, or why not, will this procedure yield random samples?

8. Suggest how you might set up an approximate random-sampling scheme for drawing samples of (a) trees in a forest, (b) potatoes in a freight car loaded with sacks of potatoes, (c) children of a community under 5 years of age who have had measles. In each case indicate some variable that might be studied.

Section 3

9. Given that x is normally distributed with mean 22 and standard deviation 4, calculate the probability that the sample mean, $\bar{x}$, based on a sample of size 64, will (a) exceed 23, (b) exceed 21.5, (c) lie between 21 and 23, (d) exceed 24.

10. Given that x is normally distributed with mean 25 and standard deviation 8, calculate the probability that the sample mean, $\bar{x}$, based on a sample of size 16, will (a) be less than 27, (b) exceed 32, (c) exceed 23, (d) be less than 20, (e) lie between 28 and 29.

11. Sketch on the same piece of paper the graph of a normal curve with mean 10 and standard deviation 2 and the graph of the corresponding mean curve for a sample of size 9.

12. What would the graph of the $\bar{x}$ curve in problem 11 have looked like if the sample size had been 36?

13. If the standard deviation of weights of first-grade children is 6 pounds, what is the probability that the mean weight of a random sample of 100 such children will differ by more than 1 pound from the mean weight for all the children?

14. A new weight producing diet is to be applied to a random sample of 25 chickens taken from a flock. If the standard deviation of the gain in weight over a one-month period is expected to be about 2 ounces, what is the probability that the mean of this sample will differ from the mean of the entire flock, if they are fed the new diet, by more than $\frac{1}{2}$ ounce?

15. The mean weight of entering male freshmen students at a certain college over the past five years is 153 pounds and the standard deviation of such weights is 20 pounds. If the mean weight of the first 100 students to register is 159 pounds, would you have reason to believe that the new freshman class is heavier than usual? Give some possible explanations.

16. Show that approximately 50 per cent of the values in a normally distributed population whose mean is 0 lie in the interval from $-.6745\sigma$ to $.6745\sigma$. The deviation $.6745\sigma$ is called the probable deviation.

Section 4

17. Have each member of the class perform the following experiment 10 times. From Table II in the appendix select 10 one-digit random numbers and calculate their mean. Bring these 10 experimental means to class, where the total set of experimental means may be classified, the histogram drawn, and the mean and standard deviation computed. These results should then be compared with theory in the same manner as in the experiment in the text. The population distribution here has $\mu = 4.5$ and $\sigma = 2.87$.

18. Verify the values of $\bar{x}$ and s given in (3) of the text.

19. Verify the values of μ and σ used in (4) of the text by deleting the decimal points in Fig. 5 and treating the resulting numbers as observed frequencies for a sample of 100 from that distribution.

20. Explain how Theorem 2 justifies the assumption made in Chapter 5 to the effect that the normal distribution is a good approximation to the binomial distribution if n is sufficiently large. Do this by considering the binomial distribution for a sample of size one in which case $x = 0$ or 1 with probabilities q and p, respectively, and using formula (7) of Chapter 5 with $x/n = \bar{x}$.

21. Perform a sampling experiment of the type used to make Theorem 2 seem plausible by taking 50 samples of size 5 each from the discrete distribution given by

x	0	1	2
$P\{x\}$	.4	.2	.4

Graph the histogram of the $\bar{x}$ distribution and calculate its mean and standard deviation. Calculate the mean and standard deviation of the x distribution and compare your $\bar{x}$ results with those expected from Theorem 2. Since n is very small here and the x distribution is far from being normal, you should not expect the $\bar{x}$ distribution to look too much like a normal distribution.

CHAPTER 7

Estimation

1. POINT AND INTERVAL ESTIMATES

The introduction in Chapter 1 stated that one of the fundamental problems of statistics is the estimation of properties of populations. Now that probability distributions have been studied, to a limited extent at least, it is possible to discuss the properties of populations that can be estimated. The two probability distributions that have been studied thus far are the binomial distribution and the normal distribution; therefore their properties will be investigated first.

The binomial distribution given by formula (1), Chapter 5, is completely determined by the number of trials, n, and the probability of success in a single trial, p. The symbols n and p are called the *parameters* of the distribution. The values assigned to the parameters determine the particular binomial distribution desired. Since the parameters n and p completely determine the binomial distribution, any property of a binomial distribution is also completely determined by them. Furthermore, since the number of trials, n, is almost always chosen in

137

advance in estimation problems, the problems of estimation for binomial distributions can usually be reduced to the problem of estimating p.

The normal distribution given by Fig. 3, Chapter 5, is completely determined by the two parameters μ and σ. Problems of estimation for normal populations can therefore usually be reduced to the problems of estimating μ and σ.

There are two types of estimates of parameters in common use in statistics. One is called a point estimate and the other is called an interval estimate. A point estimate is the familiar kind of estimate; that is, it is a number obtained from computations on the sample values that serves as an approximation to the parameter being estimated. For example, the sample proportion, x/n, of voters favoring a certain candidate is a point estimate of the population proportion p. Similarly, the sample mean $\bar{x}$ is a point estimate of the population mean μ. An interval estimate for a parameter is an interval, determined by two numbers obtained from computations on the sample values, that is expected to contain the value of the parameter in its interior. The interval estimate is usually constructed in such a manner that the probability of the interval's containing the parameter can be specified. The advantage of the interval estimate is that it shows how accurately the parameter is being estimated. If the length of the interval is very small, high accuracy has been achieved. Such interval estimates are called _confidence intervals_. Both point and interval estimates are determined for binomial and normal distribution parameters in this chapter.

2.   ESTIMATION OF μ

Consider the following problem. A manufacturer of vitamin C tablets wishes to check on the quality of his product. He has found from experience that the vitamin C content of the tablets for a given manufactured batch is approximately normally distributed. He has also found that the mean content for a batch varies from batch to batch but that the standard deviation remains fairly constant at the value $\sigma = 20$, regardless of the value of the mean. If the mean content of a batch is too low, the manufacturer cannot sell the batch; therefore it is important to him to have an accurate estimate of the batch mean. To estimate the mean content for a new batch he tests a random sample of 25 tablets taken from the production line and finds their mean vitamin C content to be $\bar{x} = 260$ units. With these data and the preceding information available, three types of estimation problems will be solved.

(a) How accurate is $\bar{x} = 260$ as a point estimate of the batch mean μ? To solve this problem, use is made of the theory presented in the preceding chapter. From the theory given in (2), Chapter 6, it follows that the sample mean $\bar{x}$ may be assumed to be normally distributed with mean μ and standard deviation given by

$$\sigma_{\bar{x}} = \frac{\sigma}{\sqrt{n}} = \frac{20}{\sqrt{25}} = 4.$$

A sketch of this distribution is shown in Fig. 1. Since the probability is .95 that a normal variable will assume some value within two standard deviations of its mean (more accurately 1.96 standard deviations correct to two decimals by Table IV in the appendix), it follows that the probability is .95 that $\bar{x}$ will assume some value within 8 units of μ. Since $\bar{x} = 260$ is the observed value here, the manufacturer can feel quite confident that this value differs from the population value μ by less than 8 units because in the long run in only 5 per cent of such sampling experiments will the sample value $\bar{x}$ differ by more than 8 units from μ. The magnitude of the difference $\bar{x} - \mu$ is called the error of estimate. In terms of this language, one can say that the probability is .95 that the error of estimate will be less than 8 units. If higher probability odds were desired, one could use, say, a three-standard deviation interval on both sides of μ and then state that the probability is .997 that the error of estimate will be less than 12 units.

Since $\bar{x}$ is based on a sample it is not possible to state how close $\bar{x}$ is to the population mean μ when μ is unknown; it is only possible to state in probability language how close $\bar{x}$ is likely to be to μ. Thus, the exact error of estimate, namely $|\bar{x} - \mu|$, is known only when μ is known. Since the problem of estimation arises only when μ is unknown, one must introduce probabilities in order to discuss

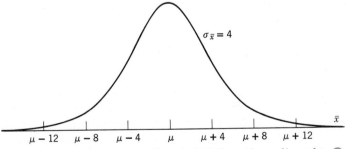

FIGURE 1. Distribution of $\bar{x}$ for the vitamin C content of tablets.

the magnitude of an error of estimate in statistical problems. Even though the true value μ is not known, one can still speak of how close $\bar{x}$ is to μ and therefore one can say something about the magnitude of the error of estimate, $|\bar{x} - \mu|$, provided the statement is couched in the proper probability language.

(b) Suppose the manufacturer is not satisfied with the accuracy of his estimate based on the sample of 25. How large an additional sample should he take so that he can be reasonably certain, say with a probability of .95, that his estimate will not be in error by more than 5 units? Now, as n is increased, the normal curve for $\bar{x}$ will become taller and narrower, concentrating more and more area in the neighborhood of the mean μ. This is shown, for example, in Figs. 3 and 4, Chapter 6, where n goes from 1 to 25. A stage will be reached when 95 per cent of the area, centered at μ, is found to lie within the interval extending 5 units on both sides of μ. This value of n is the desired sample size. Since 95 per cent of the central area corresponds to 1.96 standard deviations on both sides of μ, it follows that n must be such that 1.96 standard deviations for $\bar{x}$ equals 5. Thus n must satisfy the equation

$$1.96\sigma_{\bar{x}} = 5.$$

Since $\sigma = 20$ here, this equation is equivalent to

$$1.96\frac{20}{\sqrt{n}} = 5.$$

Solving for n gives the result $n = 61.5$. The manufacturer therefore must take an additional 37 samples, since he has taken 25 already, in order to attain the desired accuracy of estimate.

Since 1.96 is an inconvenient number to use in equations of this type, it usually suffices to replace it by 2. The solution of the equation then becomes $n = 64$. Although this approximation does yield a difference of 2 in the answer, that is hardly a large number to worry about in a total sample of 64. Furthermore, the objective here is to learn the methods of statistics and any saving of calculating energy which hopefully will be applied to thinking energy is well worth the sacrifice in accuracy of computation.

If one wants the probability that the error of estimate will not exceed 5 to be something other than .95, then it is necessary to replace the factor 1.96 (or 2) in the preceding equation by the proper z value found in Table IV corresponding to the desired probability. Thus, if the probability .90 were selected, the Table IV value of z would be 1.64. This follows from the fact that 90 per cent of the area of a normal curve lies within 1.64 standard deviations of the mean. The equation to be solved for n would then become

$$1.64 \frac{20}{\sqrt{n}} = 5.$$

The solution of this equation is $n = 43$.

Since it is bothersome to have to solve an equation like this each time, a formula which yields n more directly will be obtained. If the maximum allowable error of estimate is denoted by e and the z value corresponding to the desired probability is denoted by z_0, then the equation that must be solved for n is given by

$$z_0 \frac{\sigma}{\sqrt{n}} = e.$$

The solution of this equation, and therefore the desired formula, is given by

$$n = \frac{z_0^2 \sigma^2}{e^2}.$$

This formula enables one to determine how large a sample is needed in order to estimate μ to any desired degree of accuracy before a single sample has been taken, provided the value of σ is known. It is not necessary to have a preliminary sample available, as in the problem solved four paragraphs back. If, however, one does not know σ from other sources, nor has a good estimate of it, then it is necessary to take a preliminary sample in order to obtain an estimate of σ that can be used in the formula for determining how large n must be.

(c) Consider a third type of estimation problem for this same example. What is a 95 per cent confidence interval for μ based on the original sample of 25? If it is assumed that x is exactly normally distributed, it is clear from the theory given in (1), Chapter 6, or Fig. 1, that one can write

(1) $P\{\mu - 8 < \bar{x} < \mu + 8\} = .95.$

This is an algebraic probability statement of what was stated in geometrical language in problem (a), namely, that the probability is .95 that the point on the $\bar{x}$ axis of Fig. 1 corresponding to a sample mean $\bar{x}$ will not be more than 8 units away from the point representing the population mean μ. Now it is possible to turn this geometry, and hence the algebra, around and state that the probability is .95 that the point μ will not be more than 8 units away from the point corresponding to a sample mean $\bar{x}$. The relationship here is relative; if one point is within 8 units of a second point, then the second point will be within 8 units of the first point. This reversing of the roles of the two points will now be done algebraically by the use of inequality properties.

An inequality such as $\bar{x} < \mu + 8$ can be rearranged in the same manner as an equality, except that multiplying both sides of an inequality by a negative number will reverse the inequality sign. Thus, the inequality $2 < 5$ becomes the inequality $-2 > -5$ when it is multiplied through by -1. The inequality $\bar{x} < \mu + 8$ is seen to be equivalent to the inequality $\bar{x} - 8 < \mu$ by adding -8 to both sides of the first inequality. Similarly, $\mu - 8 < \bar{x}$ is equivalent to $\mu < \bar{x} + 8$. If these two results are combined, it will be seen that the double inequality

$$\mu - 8 < \bar{x} < \mu + 8$$

is equivalent to the double inequality

$$\bar{x} - 8 < \mu < \bar{x} + 8.$$

As a consequence, the probability statement (1) is equivalent to the probability statement

(2) $$P\{\bar{x} - 8 < \mu < \bar{x} + 8\} = .95.$$

In words, this says that the probability is .95 that the population mean μ will be contained inside the interval that extends from $\bar{x} - 8$ to $\bar{x} + 8$. This interval is written in the form $(\bar{x} - 8, \bar{x} + 8)$.

Although (1) and (2) are equivalent probability statements, they possess slightly different interpretations in terms of relative frequencies in repeated runs of this sampling experiment. For each such sampling experiment, a value of $\bar{x}$ is obtained. If these values of $\bar{x}$ are plotted as points, as shown in Fig. 2, then the frequency

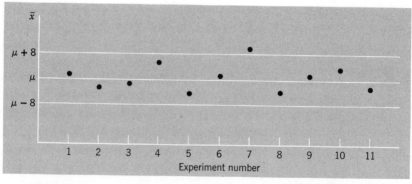

FIGURE 2.   Repeated sampling experiments for $\bar{x}$.

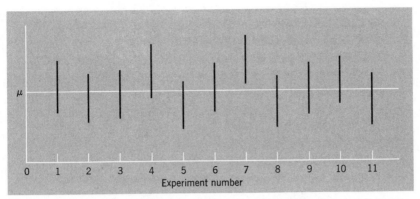

FIGURE 3.   Intervals for repeated sampling experiments.

interpretation of (1) is that in such repeated sampling experiments 95 per cent of the points will fall within the band shown in Fig. 2.

A frequency interpretation for (2) requires that the interval extending from $\bar{x} - 8$ to $\bar{x} + 8$, corresponding to each sampling experiment, be plotted. This has been done in Fig. 3 for the experiments that yielded Fig. 2. The frequency interpretation of (2), then, is that in such repeated sampling experiments 95 per cent of the intervals will contain μ. Geometrically, it is clear from Figs. 2 and 3 that an interval in Fig. 3 will contain μ if, and only if, the corresponding point in Fig. 2 lies inside the band displayed there. This is very much like saying that a chalk line on the floor (μ) will be within 8 feet of you ($\bar{x}$) if, and only if, you are within 8 feet of the line. The advantage of the interval interpretation is that in practice one never knows what the value of μ is; otherwise there would be no point of estimating it, and therefore it is not possible to construct the band given by $\mu - 8$ and $\mu + 8$ in Fig. 2; however, it is always possible to construct the intervals given by $\bar{x} - 8$ and $\bar{x} + 8$ in Fig. 3.

Now, in practice, only one sampling experiment is conducted; therefore, only the first point and the first interval are available from Figs. 2 and 3. On the basis of this one experiment, the claim is made that the interval from $260 - 8$ to $260 + 8$, or from 252 to 268, contains the population mean μ. Using inequality symbols, this is written in the form

$$252 < \mu < 268.$$

If for each such sampling experiment the same claim is made for the interval corresponding to that experiment, then 95 per cent of such claims will be true in the long run of such experiments. In view of this property, the interval from

252 to 268 is called a 95 per cent *confidence interval* for μ. The end points of the interval, namely 252 and 268, are called *confidence limits* for μ.

It should be clearly understood that one is merely betting on the correctness of the rule of procedure when applying the confidence interval technique to a given experiment. It is incorrect to make the claim that the probability is .95 that the interval from 252 to 268 will contain μ. The latter probability is either 1 or 0, depending upon whether μ does or does not lie in this fixed interval. Nontrivial probability statements are made only about variables and not about constants. It is only when one considers the variable interval from $\bar{x} - 8$ to $\bar{x} + 8$, before a numerical value of $\bar{x}$ has been obtained, that one can make probability statements such as that in (2).

The advantage of a confidence interval for μ over a point estimate of μ is that the confidence interval gives one an idea of how closely μ is being estimated, whereas the point estimate $\bar{x}$ says nothing about how good the estimate is. Thus, the confidence interval $(252, 268)$ gives one assurance (confidence) that the true mean μ is very likely at least as large as 252 and very likely not larger than 268.

The three types of problems just solved in connection with the example introduced at the beginning of this section are the three major problems arising in the estimation of μ. They may be listed as (a) determining the accuracy of $\bar{x}$ as an estimate of μ, (b) determining the size sample needed to attain a desired accuracy of estimate of μ, and (c) determining a confidence interval for μ.

The methods for solving these problems were quite simple because it was assumed that the variable x was approximately normally distributed and that the value of σ was known. Now, by the theory in (2), Chapter 6, it follows that it would have been safe to treat $\bar{x}$ as a normal variable, even though x had not been assumed to be normally distributed, because n is large here. The problem of what to do when σ is not known is not so simple. If the sample is large, say 25 or more, it is usually safe to replace σ by its sample estimate s in the formulas used to solve the problems.

As an illustration of what to do when σ is not known, the following problem will be worked. A random sample of 100 students is selected from a certain school. They are given an intelligence test to determine their intelligence quotient scores. The scores on this test yielded the sample values $\bar{x} = 112$ and $s = 11$. What is a 95 per cent confidence interval for the school mean intelligence quotient based on these sample values? The problem is solved in the same manner as before, except that s is used in place of σ.

As before, it follows from Table IV that the probability is .95 that a standard normal variable z will satisfy the inequalities

$$-1.96 < z < 1.96.$$

But if $\bar{x}$ is a normal variable, the quantity

$$\frac{\bar{x} - \mu}{\sigma_{\bar{x}}} = \frac{\bar{x} - \mu}{\sigma/\sqrt{n}} = \frac{\bar{x} - \mu}{\sigma}\sqrt{n}$$

will be a standard normal variable, and therefore the probability is .95 that it will satisfy the inequalities

$$-1.96 < \frac{\bar{x} - \mu}{\sigma}\sqrt{n} < 1.96.$$

If these inequalities are solved for μ, they will reduce to the following inequalities:

(3)
$$\bar{x} - 1.96\frac{\sigma}{\sqrt{n}} < \mu < \bar{x} + 1.96\frac{\sigma}{\sqrt{n}}.$$

This result can be used as a formula for obtaining 95 per cent confidence intervals for population means.

For the problem being considered, $n = 100$, $\bar{x} = 112$, and $s = 11$. Since the value of σ is not known here, it must be approximated by its sample estimate $s = 11$. If these values are substituted into (3), it will assume the form

$$112 - 1.96\frac{11}{\sqrt{100}} < \mu < 112 + 1.96\frac{11}{\sqrt{100}}.$$

These quantities reduce to 109.8 and 114.2; hence the desired approximate 95 per cent confidence interval for μ is given by

$$109.8 < \mu < 114.2.$$

This is only an approximate 95 per cent confidence interval because σ was replaced by its sample approximation s and x was not assumed to be normally distributed. For a sample as large as 100, the errors arising because of these approximations will be negligible.

If one desired, say, a 90 per cent confidence interval rather than a 95 per cent confidence interval, it would merely be necessary to replace the number 1.96 by the number 1.64 in the preceding formulas, just as in the earlier problem of determining n. In the preceding problem the limits would then become

$$112 - 1.64\frac{11}{\sqrt{100}} < \mu < 112 + 1.64\frac{11}{\sqrt{100}}.$$

If these inequalities are simplified, the desired approximate 90 per cent confidence interval for μ will become

$$110.2 < \mu < 113.8.$$

Any other percentage confidence interval can be obtained in a similar manner by means of Table IV.

The methods of estimation explained in this section are called large sample methods whenever σ is replaced by its sample estimate because they are then strictly valid only for large samples. If the sample is smaller than about 25 and the value of σ is unknown, these methods are of questionable accuracy, and therefore, a more refined method is needed. A method designed to solve such small sample problems is presented next.

3. STUDENT'S t DISTRIBUTION

Consider once more the problem that was solved three paragraphs back, with the modification that the sample size is given to be 10 rather than 100. Thus, $\bar{x} = 112$ and $s = 11$ for a sample of size 10. To avoid the error involved in replacing σ by s when s is based on such a small sample, a new variable, called *Student's t variable* is introduced. It is defined by the formula

(4)
$$t = \frac{\bar{x} - \mu}{s} \sqrt{n}.$$

This variable resembles the standard normal variable introduced in section 2, namely,

$$z = \frac{\bar{x} - \mu}{\sigma} \sqrt{n}.$$

However, it differs from z in that it involves the sample standard deviation s in place of the population standard deviation σ. Since t does not require a knowledge of σ, as is the case with z, its value can be computed from sample data, whereas the value of z cannot be computed unless σ is known. This is the reason why t can be used to solve problems without the necessity of introducing approximations to population parameters.

If a large number of sampling experiments were carried out in which a sample of size n were selected from a normal population and the value of t computed, a large number of values of t would be available for classifying into a frequency table to obtain a good estimate of the limiting, or theoretical, distribution of t. Mathematical methods, however, yield the exact distribution. It turns out that the distribution of t depends only upon the value of n, provided that the basic variable x possesses a normal distribution. Furthermore, the distribution of t is

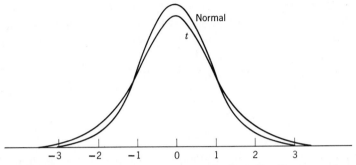

Normal

t

−3 −2 −1 0 1 2 3

FIGURE 4. Standard normal distribution and a Student's t distribution when n = 5.

very close to the distribution of a standard normal variable z, except for very small values of n. Figure 4 shows the graph of the distribution of t for $n = 5$ and the graph of a standard normal variable z.

Table V in the appendix gives values of the variable t corresponding to what is called the number of "degrees of freedom," denoted by v, and various probabilities. For the problem being considered here, the number of degrees of freedom is given by the formula $v = n - 1$. This corresponds to using the divisor $n - 1$ rather than n in defining the sample standard deviation in Chapter 2. The t distribution is used for other types of problems also in which the parameter v is not equal to $n - 1$; otherwise this mysterious phrase would not need to be introduced here. Any column heading, such as .025, indicates the probability of t exceeding the value of t listed in that column. Because of symmetry it follows that .025 is also the probability that t will lie to the left of the corresponding $-t$ value. Thus for the above problem, since $n = 10$, one reads the entry in the row corresponding to 9 degrees of freedom and in the column headed .025, and finds $t = 2.26$. The probability is therefore .95 that t will satisfy the inequalities

(5) $$-2.26 < t < 2.26.$$

If this is applied to (4), it will follow that the probability is .95 that

$$-2.26 < \frac{\bar{x} - \mu}{s} \sqrt{n} < 2.26.$$

These inequalities can be solved for μ in the same manner as for formula (3). The result is

(6) $$\bar{x} - 2.26\frac{s}{\sqrt{n}} < \mu < \bar{x} + 2.26\frac{s}{\sqrt{n}}.$$

The desired 95 per cent confidence interval for μ is obtained by substituting the sample values $n = 10$, $\bar{x} = 112$, and $s = 11$ into these inequalities. The result of this substitution is

$$112 - 2.26\frac{11}{\sqrt{10}} < \mu < 112 + 2.26\frac{11}{\sqrt{10}}.$$

This simplifies into

(7) $104 < \mu < 120.$

It is important to realize the distinction between this new method of finding a confidence interval for μ and the earlier large sample method. This method does not require one to approximate σ by s, as is true for the large sample method, and therefore it gives an exact rather than an approximate solution to the problem.

Formula (6) cannot be used to find a 95 per cent confidence interval for μ unless the sample size is 10, because from Table V it will be found that (5) holds only for $\nu = 9$. If t_0 is used to denote the value of t found in the .025 column of Table V opposite ν degrees of freedom, then a general formula for a 95 per cent confidence interval for μ is given by the inequalities

(8) $\bar{x} - t_0\frac{s}{\sqrt{n}} < \mu < \bar{x} + t_0\frac{s}{\sqrt{n}}.$

This formula is also valid for percentages other than 95 if the corresponding value of t_0 is employed. Thus a 90 per cent confidence interval is obtained if one replaces t_0 by the t value in the .05 column of Table V, which is opposite the desired degrees of freedom value.

For the purpose of comparing the old method with the new, it is necessary to calculate a 95 per cent confidence interval by means of formula (3). If σ is estimated by s and the sample values $n = 10$, $\bar{x} = 112$, and $s = 11$ are substituted in (3), the desired approximate 95 per cent confidence interval is given by

$$112 - 1.96\frac{11}{\sqrt{10}} < \mu < 112 + 1.96\frac{11}{\sqrt{10}}.$$

This simplifies into

$$105 < \mu < 119.$$

It will be noted that this interval is somewhat narrower than that given by the small sample method and displayed in (7). The large sample method always gives a confidence interval that is somewhat too narrow; however, the error decreases rapidly as n grows and is hardly noticeable for n larger than 20.

Since the small sample method based on the t distribution is an exact method, it would seem that one should always use it when σ is unknown. However, the theory behind the t distribution requires one to assume that the basic variable x possesses a normal distribution; therefore unless one can be assured that x is at least approximately normally distributed, the t distribution may not be very accurate. The large sample method requires only that $\bar{x}$ be normally distributed and that s be a good estimate of σ. Now $\bar{x}$ will be approximately normally distributed even though x is not, even for fairly small samples, as was shown by the sampling experiments in the preceding chapter. For non-normal variables the large sample method would therefore appear to be more appropriate. There is, however, little difference in the two methods then.

4.	ESTIMATION OF p

Section 2 was concerned with the estimation of the parameter μ for continuous variable distributions. This section explains how to solve similar types of problems for the parameter p associated with binomial distributions. The methods presented here are large sample methods because they require the replacement of p by its sample estimate and also because they assume that the normal curve approximation to the binomial distribution is satisfactory.

As an example to illustrate the various types of estimation problems to be solved, consider the problem of estimating the percentage of adult males in a certain city who smoke at least one pack of cigarettes a day. Suppose a random sample of size 300 yielded 36 such individuals. By using these data, the following three problems can be solved: (a) What is the accuracy of the sample proportion as an estimate of p? (b) How large a sample would be needed if the probability is to be .95 that the error of estimate will not exceed .02 units? (c) What is a 95 per cent confidence interval for p? All of these problems are solved in the same manner as in section 2 because the sample size here is large enough to justify the use of normal curve methods.

(a) From formula (7), Chapter 5, it follows that the sample proportion, x/n, which will be denoted by $\hat{p}$, may be assumed to be approximately normally distributed with mean p and standard deviation

$$\sqrt{\frac{pq}{n}} = \sqrt{\frac{pq}{300}}.$$

As a result, the probability is approximately .95 that $\hat{p}$ will lie within 1.96 such standard deviations of p. Thus the probability is approximately .95 that the error of estimate will be less than

(9)
$$1.96 \sqrt{\frac{pq}{300}}.$$

Since p is unknown, it must be estimated by

$$\hat{p} = \frac{x}{n} = \frac{36}{300} = .12.$$

The value of (9) then assumes the approximate value

$$1.96 \sqrt{\frac{(.12)(.88)}{300}} = .037.$$

It can therefore be stated that the probability is approximately .95 that the sample estimate $\hat{p}$ will not differ from p by more than .037 units. This result gives one a good idea of the accuracy of the sample value .12 as an estimate of p.

(b) To solve the problem of how large a sample is needed to attain a given accuracy of estimate for p, one uses the same reasoning as that used in section 2 for μ. This means that n must be chosen so that the proper number of standard deviations of $\hat{p}$ will equal the desired maximum error of estimate. As before, let e denote the selected maximum error of estimate and let z_0 denote the value of z corresponding to the desired probability of not exceeding this maximum error. Then n must satisfy the equation

$$z_0 \sqrt{\frac{pq}{n}} = e.$$

Solving this equation for n will yield the formula

(10)
$$n = \frac{z_0^2 pq}{e^2}.$$

For the particular problem being considered here, $e = .02$ and $z_0 = 1.96$. Since p is unknown, it must be estimated by the sample value $\hat{p} = .12$. If these values are substituted in (10), the value of n will be found to be approximately 1014; hence an additional sample of approximately 714 will be needed to obtain the desired accuracy of estimation.

(c) To find a confidence interval for p, one also uses the same reasoning as for μ. Since $\hat{p}$ takes the place of $\bar{x}$, an approximate 95 per cent confidence interval for p is given by the inequalities

(11)
$$\hat{p} - 1.96\sqrt{\frac{pq}{n}} < p < \hat{p} + 1.96\sqrt{\frac{pq}{n}}.$$

Replacing p by $\hat{p}$ in the two radical terms and substituting the values of $n = 300$ and $\hat{p} = .12$, one obtains the approximate interval

$$.12 - .037 < p < .12 + .037.$$

These limits reduce to .083 and .157; consequently an approximate 95 per cent confidence interval for p is given by the inequalities

$$.083 < p < .157.$$

Repetition may be boring, yet it is worth repeating that all three solutions are based on large sample methods. Fortunately, these methods are quite good, even for small samples, provided that $np > 5$ for $p < \frac{1}{2}$ and $nq > 5$ for $p > \frac{1}{2}$. Better methods are available for small values of n but they are not treated here.

A final illustration of the methods for estimating p is given because of its interest to those who enjoy politics. A well known pollster claims that his estimate of the proportion of the voters favoring a certain presidential candidate is not in error by more than .03 units. In a close presidential race, how large a sample would he need to take to be certain, with a probability of .997, of being correct in his claim?

From Table IV in the appendix, 99.7 per cent of the central area of a normal distribution lies within three standard deviations of the mean; therefore $z_0 = 3$ here. Since the race is very close, it may be assumed that $p = \frac{1}{2}$; hence formula (10) yields the result

$$n = \frac{9 \cdot \frac{1}{2} \cdot \frac{1}{2}}{(.03)^2} = 2500.$$

A random sample of this size taken from over the country should therefore suffice to give him the desired accuracy.

The use of $p = \frac{1}{2}$ in the foregoing problem may appear to be arbitrary, particularly if the election is not really close. However, it is easy to show that pq assumes its maximum value when $p = \frac{1}{2}$, and hence that the maximum value of n in (10) occurs when $p = \frac{1}{2}$. This is done by first verifying that

$$pq = p(1-p) = p - p^2 = \frac{1}{4} - \left(\frac{1}{2} - p\right)^2.$$

Next, pq will be as large as possible, namely $\frac{1}{4}$, when the term $(\frac{1}{2} - p)^2$ which is being subtracted from $\frac{1}{4}$ has the value 0. But this will occur when $p = \frac{1}{2}$. This implies that when one is determining the size of the sample necessary for a specified accuracy of estimate, the value of n for $p = \frac{1}{2}$ will be larger than for any other value of p. As a result, the use of $p = \frac{1}{2}$ in such problems assures one that the resulting value of n is certainly large enough and possibly larger than necessary.

An interesting feature of problems like this one is that, contrary to the belief of most people, the accuracy of an estimate of a proportion p does not depend upon the size of the population but only upon the size of the sample. Thus a sample of 2500 voters out of 50,000,000 voters is sufficient, theoretically, to determine their voting preferences with high accuracy.

Unfortunately, voters do not always behave like trials in a game of chance, so that the binomial distribution model is not strictly applicable to voting problems. For example, a voter when interviewed may favor one candidate and yet a week later he may vote for another candidate, or he may not bother to vote at all. He may also misinterpret a pollster's question and therefore respond incorrectly. Experience has shown that because of uncontrolled human factors the accuracy of an estimate of p for voters does not increase appreciably after a sample of 10,000 has been taken. It is necessary to use good sense in applying mathematical models to real life, particularly when it comes to human beings and some of their inconsistencies. Wild animals seem to be better subjects on which to apply statistical methods.

5.   ADDITIONAL ILLUSTRATIONS

1. A gasoline additive is being tested to see whether it increases mileage. Twenty-five cars are supplied with 5 gallons of gasoline and are run until the gasoline is exhausted. At the completion of the experiment the average mileage for each car is computed. Calculations with the data of this one experiment gave a mean of $\bar{x} = 18.5$ miles per gallon and a standard deviation of $s = 2.2$ miles per gallon for the 25 cars. Experience with cars of the same kind that were used before when no additive was employed indicates that, approximately, $\mu = 18.0$ and $\sigma = 2.0$ miles per gallon. Assuming that the additive had no effect on mileage, solve the following problems. (a) 2 Determine the probability accuracy of $\bar{x}$ as an estimate of μ. What is the actual accuracy? Is the sample compatible with what was to be expected by theory? (b) 2 Find how large an experiment should

have been conducted if one wished to be certain with a probability of .95 that
the estimate would not be in error by more than $\frac{1}{2}$ mile per gallon. (c) 2 Find
a 95 per cent confidence interval for μ. Does this interval actually contain μ? (d)
3 Dropping the assumption that the additive had no effect on either the mean
or variance, use student's t variable to find a 95 per cent confidence interval for
μ. The solutions follow.

(a) Here $\mu = 18.0$, $\sigma = 2.0$, $n = 25$, and $\bar{x} = 18.5$; hence

$$1.96 \, \sigma_{\bar{x}} = 1.96 \, \frac{2.0}{\sqrt{25}} = .784.$$

The probability is .95 that the error of estimate will not exceed .784. The actual
error is $|\bar{x} - \mu| = 18.5 - 18.0 = .5$; therefore the sample value is compatible
with theory.

(b) Since .95 gives $z_0 = 1.96$ and the maximum tolerable error is to be $e = \frac{1}{2}$,
n is given by

$$n = \frac{(1.96)^2(2.0)^2}{(\frac{1}{2})^2} = 61.5.$$

An additional sample of 37 would suffice. Thus, approximately 37 additional
cars of the same type should be run.

(c) A 95 per cent confidence interval is given by

$$\bar{x} - 1.96 \frac{\sigma}{\sqrt{n}} < \mu < \bar{x} + 1.96 \frac{\sigma}{\sqrt{n}}.$$

Hence, for this problem it becomes

$$18.5 - 1.96 \frac{2.0}{\sqrt{25}} < \mu < 18.5 + 1.96 \frac{2.0}{\sqrt{25}}$$

Or,

$$17.72 < \mu < 19.28.$$

Since $\mu = 18.0$, it is contained inside this interval.

(d) Since $\nu = 24$ and a 95 per cent interval is desired, $t_0 = 2.0639$; hence
formula (8) gives

$$18.5 - 2.06 \frac{2.2}{\sqrt{25}} < \mu < 18.5 + 2.06 \frac{2.2}{\sqrt{25}}.$$

Or,

$$17.59 < \mu < 19.41.$$

This interval also contains $\mu = 18.0$; therefore it is compatible with the assumption of no additive effect.

2. Work parts (b) 2, (c) 2, and (d) 3 of the preceding problem using a probability of .90 in place of .95.

(b) Since $z_0 = 1.64$,

$$n = \frac{(1.64)^2(2.0)^2}{(\frac{1}{2})^2} = 43.0.$$

An additional sample of 18 is needed.

(c) Here one uses 1.64 in place of 1.96 to obtain

$$18.5 - 1.64\frac{2.0}{\sqrt{25}} < \mu < 18.5 + 1.64\frac{2.0}{\sqrt{25}}.$$

Or,

$$17.84 < \mu < 19.16.$$

(d) Here $t_0 = 1.7109$; hence the interval is given by

$$18.5 - 1.71\frac{2.2}{\sqrt{25}} < \mu < 18.5 + 1.71\frac{2.2}{\sqrt{25}}.$$

Or,

$$17.75 < \mu < 19.25.$$

3. An insurance company has found from experience that 30 per cent of its insured have had at least one automobile accident during the past three years. It is contemplating insuring all the civil service employees of a certain city and wishes to determine their insurance rates. It takes a sample of 100 of those employees and discovers that 25 of them have had at least one accident during the past three years. Assuming that these employees are typical of the company's insured, work the following problems. (a) 4 What is the probability accuracy of this estimate? What is the actual accuracy? Is the sample proportion compatible with what was to be expected by theory? (b) 4 Find how large an additional sample would be needed if one wished to estimate p to within .03 units with a probability of .95 of being correct. Assume that p is not known here. (c) 4 Work part (b) if the conservative value $p = \frac{1}{2}$ is used in place of the sample estimate of p. (d) 4 Find a 90 per cent confidence interval for p, assuming that the value of p is unknown. Does this interval actually contain p? The solutions follow.

(a) Here $\hat{p} = .25$. Since $p = .30$ and $n = 100$,

$$1.96\,\sigma_{\hat{p}} = 1.96\sqrt{\frac{(.30)(.70)}{100}} = .09.$$

The probability is .95 that the error will not exceed .09. The actual error is $|\hat{p} - p| = |.25 - .30| = .05$; therefore the sample value is compatible with theory.

(b) Since $z_0 = 1.96$ and $e = .03$, n is given by

$$n = \frac{(1.96)^2 pq}{(.03)^2} \doteq \frac{(1.96)^2(.25)(.75)}{(.03)^2} = 800.$$

← why not .30 ?

Thus, 700 additional samples would be required.

(c) Using $p = \frac{1}{2}$,

$$n = \frac{(1.96)^2(\frac{1}{2})(\frac{1}{2})}{(.03)^2} = 1067.$$

Thus, 967 additional samples would be needed. Using this very conservative value of p would increase the cost of sampling considerably here.

(d) Using $\hat{p} = .25$ in place of p under the radical of formula (11) gives

$$.25 - 1.64 \sqrt{\frac{(.25)(.75)}{100}} < p < .25 + 1.64 \sqrt{\frac{(.25)(.75)}{100}}.$$

Or

$$.18 < p < .32.$$

The value $p = .30$ is inside this interval.

4. Work parts (b) 4 and (d) 4 of the preceding problem using a probability of .80 in each part. The solutions follow.

(b) Here $z_0 = 1.28$; hence

$$n = \frac{(1.28)^2(.25)(.75)}{(.03)^2} = 341.$$

Thus, 241 additional samples will suffice.

(d) Here

$$.25 - 1.28 \sqrt{\frac{(.25)(.75)}{100}} < p < .25 + 1.28 \sqrt{\frac{(.25)(.75)}{100}}.$$

Or

$$.195 < p < .305.$$

Now the value $p = .30$ is barely inside the confidence interval.

EXERCISES

Section 2

1. Experience with workmen in a certain industry indicates that the time required for a randomly selected workman to complete a job is approximately normally distributed with a standard deviation of 12 minutes. (a) If each of a random sample of 16 workmen performed the job, how accurate is their sample mean as an estimate of the mean for all the workmen? (b) How much improvement would have resulted in the accuracy of this estimate if 64 workmen had been selected?

2. From past experience the standard deviation of the height of fifth-grade children in a school system is 2 inches. (a) If a random sample of 25 such children is taken, how accurate would their sample mean be as an estimate of the mean for all such children? (b) What would happen to the accuracy of this estimate if the sample were made 4 times as large?

3. (a) In problem 1(a) how large a sample would one need to take if one wished to estimate the population mean to within 3 minutes, with a probability of .95 of being correct? (b) What size sample would be needed if the maximum error of estimate were to be 1 minute?

4. Work problem 3 for the case in which one is satisfied to have a probability of .90 of being correct.

5. (a) In problem 2 how large a sample would one need to take if one wished to estimate the population mean to within $\frac{1}{2}$ inch, with a probability of .95 of being correct? (b) What size sample would be needed if the maximum error of estimate were to be $\frac{1}{5}$ inch?

6. Work problem 5(a) for the case in which one is satisfied to have a probability of .90 of being correct.

7. If the results of the experiment in problem 1(a) yielded $\bar{x} = 140$ minutes, find (a) 95 per cent confidence limits for μ, (b) 90 per cent confidence limits for μ.

8. If the results of the experiment in problem 2(a) yielded $\bar{x} = 54$ inches, find (a) 95 per cent confidence limits for μ, (b) 90 per cent confidence limits for μ.

9. A set of 50 experimental animals is fed a certain kind of rations for a 2-week period. Their gains in weight yielded the values $\bar{x} = 42$ ounces and $s = 6$ ounces. (a) How accurate is 42 as an estimate of the population mean? (b) How large a sample would you take if you wished $\bar{x}$ to differ from μ by less than 1 ounce, with a probability of .95 of being correct? (c) Find 95 per cent confidence limits for μ. Use large sample methods here.

10. If a sample of 100 has been taken and $\bar{x} = 40$, $s = 8$ resulted, with what probability can one be assured that $\bar{x}$ is not more than 1 unit away from the true mean?

11. A chicken farmer has a flock of 1000 chickens. He wishes to experiment with a new weight producing diet. If he knows that the standard deviation of weight gain over a one-month period is 2 ounces, how large a sample should he take for experimental purposes so that his estimate of the total gain in weight for the 1000 chickens, if placed on this diet, will, with a probability of .95, not be in error by more than 40 pounds?

12. Suggest how you might proceed to determine the sample size needed for estimating μ with a certain accuracy when σ is unknown by taking samples in small groups and re-estimating σ as additional groups are taken.

13. Have each member of the class find a 75 per cent confidence interval for μ for a sample of size 25 from a table of one-digit random numbers (Table II in the appendix). Use the fact that σ for this distribution is given by $\sigma = 2.87$ and that $\bar{x}$ may be treated as a normal variable. Check to see what percentage of the student's confidence intervals contain the true mean $\mu = 4.5$. About 75 per cent should do so.

Section 3

14. Given that $\bar{x} = 20$, $s = 5$, $n = 10$, with x normally distributed, use Student's t distribution to find (a) 95 per cent confidence limits for μ, (b) 99 per cent confidence limits for μ.

15. A sample of 20 cigarettes of a certain brand was tested for nicotine content and gave $\bar{x} = 22$ and $s = 4$ milligrams. Use Student's t distribution to find 95 per cent confidence limits for μ.

16. Work problem 15 by large sample methods and compare the results of the two methods.

17. A set of 12 experimental animals was fed a special diet for 3 weeks and produced the following gains in weight: 30, 22, 32, 26, 24, 40, 34, 36, 32, 33, 28, 30. Find 90 per cent confidence limits for μ.

18. Work problem 13, but this time use a sample of size 10 and assume that the value of σ is not known. That is, use Student's t distribution to find the desired confidence interval. Check to see what percentage of the students' intervals contain μ.

19. A sleep producing drug was administered to 20 patients and produced a mean increase in sleep of 32 minutes. If the standard deviation for the 20 patients was 18 minutes, find a 90 per cent confidence interval for the mean increase for patients of the type being treated.

20. In problem 19 find the confidence intervals that would have resulted if the sample had been one of size (a) 10, (b) 40. On the basis of these two results, what can be said about the width of the confidence interval as a function of the sample size?

Section 4

21. A campus organization wishes to estimate the percentage of students who favor a new student body constitution. It proposes to select a random sample of 300 students. If the results of this poll yield $\hat{p} = .60$, how accurate is this estimate of the true proportion likely to be?

22. A sample of 60 motorists showed that 20 per cent had lapsed driver's licenses. How accurate is this estimate of the true percentage likely to be?

23. If an estimate, accurate to within .04 units, is desired of p in problem 21, how large a sample should the organization plan to take? Assume that a probability of .95 of being correct will suffice and use $\hat{p} = .60$.

24. A manufacturer of parts believes that approximately 4 per cent of his product contains flaws. If he wishes to estimate the true percentage to within $\frac{1}{2}$ per cent and to be certain with a probability of .99 of being correct, how large a sample should he take?

25. If the campus organization in problem 23 had no experience to give it the estimate $\hat{p} = .60$, how large a sample should it plan on taking?

26. A random sample of 400 citizens in a community showed that 280 favored having their water fluoridated. Use these data to find 95 per cent confidence limits for the proportion of the population favoring fluoridation.

27. A campus organization wished to estimate the percentage of the student body favoring their candidate in a forthcoming election. They proceeded to do so by asking the first 200 students whom they met going to 8 o'clock classes for their views. They found that 30 per cent favored their candidate. (a) Criticize the validity of this estimate. (b) If the estimate were valid, what would 96 per cent confidence limits be for p?

28. Obtain 80 per cent confidence limits for the number of accident claims that will be paid by an insurance company during the next year if this year's experience showed that 6 per cent of those carrying insurance collected claims and the company has 6000 policies.

29. Solve each of the inequalities in the double inequality $\hat{p} - z_0\sqrt{pq/n} < p < \hat{p} + z_0\sqrt{pq/n}$ for the variable p. This will involve the solution of a quadratic equation. Use your results to obtain a confidence interval for p that does not contain p in its limits.

CHAPTER 8

Testing Hypotheses

1. TWO TYPES OF ERROR

As indicated in Chapter 1, a second fundamental problem of statistics is the testing of hypotheses about populations. From the discussion on estimation in Chapter 7, it follows that the testing of hypotheses about binomial populations can usually be reduced to testing some hypothesis about the parameter p. Similarly, the testing of hypotheses about normal populations can usually be reduced to testing hypotheses about the parameters μ and σ.

Examples, which are essentially hypothesis testing problems related to binomial or normal distributions, have already been discussed. For example, the problem of determining by means of a sample of 400 voters whether a politician's claim of 60 per cent backing was valid is a problem of testing the hypothesis that $p = .6$ for a binomial distribution for which $n = 400$. The problem of comparing weights of dormitory and nondormitory students, introduced in Chapter 2, can be treated as a problem of testing the hypothesis that the means and standard deviations of two normal distributions are equal.

159

For the purpose of explaining the methods used to test a hypothesis about a population parameter, consider a particular problem.

During the last fifty years or more, archaeologists in a certain country have been attempting to classify skulls found in excavations into one of two racial groups, partly by the pottery and other utensils found with the skulls and partly by differences in skull dimensions. In particular, they have found that the mean length of all the skulls found thus far from race A is 190 millimeters, whereas the mean length of those from race B is 196 millimeters. The standard deviation of such measurements of length was found to be about the same for the two groups and approximately equal to 8 millimeters. A new excavation produced 12 skulls, which there is reason to believe belong to race A. The mean length of these skulls is $\bar{x} = 194$ millimeters. The problem is to test the hypothesis that the skulls belong to race A rather than to race B.

Since the test is to be based on the value of $\bar{x}$, it is formulated as a test of the hypothesis, denoted by H_0, that the population mean for the 12 skulls is 190, as contrasted to the alternative hypothesis, denoted by H_1, that the population mean is 196. This can be condensed as follows:

(1)
$$H_0 : \mu = 190$$
$$H_1 : \mu = 196.$$

There are two possibilities for making the wrong decision here. If the skulls really belong to race A and on the basis of the value of $\bar{x}$ one decides to accept H_1, an incorrect decision will be made. If, however, the skulls really belong to race B and one decides to accept H_0, an incorrect decision will also be made. The first type of wrong decision is usually called the type I error, whereas the second type of wrong decision is called the type II error. These two possibilities of incorrect decisions, together with the two possibilities for correct decisions, are listed in Table 1.

TABLE 1

	H_0 True	H_1 True
H_0 accepted	correct decision	type II error
H_1 accepted	type I error	correct decision

Now most people would use good sense in this particular problem and decide in favor of H_0 if $\bar{x}$ were closer to 190 than to 196 and in favor of H_1 if the reverse were true. Thus most people would accept H_1 in this problem. However, archaeologists who have other reasons for believing that the skulls belong to race A, such as pieces of pottery found with the skulls, would not be willing to use the halfway point between the two means as the borderline value for making decisions based on $\bar{x}$. They would undoubtedly insist that $\bar{x}$ be fairly close to the mean corresponding to H_1 before they would be willing to give up the hypothesis H_0 in favor of H_1. To study the reasonableness of using the halfway point, and other points to the right of it for making decisions, the probabilities of making the two types of error are calculated.

For the purpose of calculating these probabilities, it is assumed that x, the length of a skull, is approximately normally distributed with standard deviation $\sigma = 8$ and with mean $\mu = 190$, if the skull is from race A, and with mean $\mu = 196$, if the skull is from race B. Then $\bar{x}$ may be assumed to be normally distributed with standard deviation

$$\sigma_{\bar{x}} = \frac{\sigma}{\sqrt{n}} = \frac{8}{\sqrt{12}} = 2.31$$

and with mean 190 if the skulls are from race A and with mean 196 if they are from race B. The graphs of the two normal curves for $\bar{x}$ corresponding to H_0 and H_1 are shown in Fig. 1.

If the halfway point, 193, is used for the borderline of decisions, then the probability of making a type I error, that is, the probability of accepting H_1 when H_0 is true, is the probability that $\bar{x} > 193$ when H_0 is true. This probability is

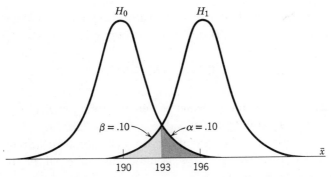

FIGURE 1.   Distribution of $\bar{x}$ under H_0 and H_1.

equal to the shaded area *under the H_0 curve* to the right of $\bar{x} = 193$. Its value, which is denoted by α, was found by the methods explained before to be .10. The probability of making a type II error, that is, the probability of accepting H_0 when H_1 is true, is the probability that $\bar{x} < 193$ when H_1 is true. This probability is equal to the shaded area *under the H_1 curve* to the left of $\bar{x} = 193$. Its value, which is denoted by β, is, by symmetry, the same as that for α; hence $\alpha = .10$ and $\beta = .10$.

If an archaeologist is fairly confident, through other sources of information, that the hypothesis H_0 is true, he will wish to make the probability of rejecting H_0 when it is actually true considerably smaller than the probability of rejecting H_1 when it is actually true. Thus he will want α to be considerably smaller than β. Now it is clear from Fig. 1 that if a point to the right of 193 were chosen for the borderline of decisions the value of α would become smaller than .10 and the value of β would become larger than .10. Since it is not possible to decrease α without increasing β, the archaeologist will need to show some constraint in decreasing α or he will be faced with an unbearably large value of β. Suppose he decides that a value of $\alpha = .05$ will be small enough to give him the protection he desires against incorrectly rejecting H_0. This means that in only about one experiment in twenty will he incorrectly reject H_0 when it is true. With this choice agreed upon, it becomes necessary to select a value of $\bar{x}$ to the right of the halfway point such that the probability of making a type I error will be equal to $\alpha = .05$. Now, from Table IV in the appendix, it is known that 5 per cent of the area of the standard normal curve lies to the right of $z = 1.64$. Since $\mu = 190$ and $\sigma_{\bar{x}} = 2.31$ here and since

$$z = \frac{\bar{x} - \mu}{\sigma_{\bar{x}}},$$

is a standard normal variable, it follows that the value of $\bar{x}$ that cuts off a 5 per cent right tail of the $\bar{x}$ curve is obtained by solving for $\bar{x}$ in the equation

$$1.64 = \frac{\bar{x} - 190}{2.31}.$$

The solution of this equation, which is denoted by $\bar{x}_0$, is given by $\bar{x}_0 = 193.8$. Another manner of arriving at this value is to argue that it is necessary to go 1.64 standard deviations to the right of the mean of a normal distribution to obtain a value such that 5 per cent of the area under the curve will be to the right of it. Thus, $\bar{x}_0$ must be given by

$$\bar{x}_0 = 190 + 1.64(2.31) = 193.8$$

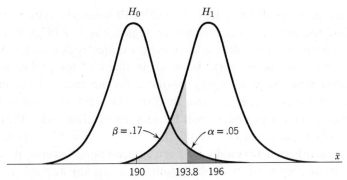

FIGURE 2. Distribution of $\bar{x}$ under H_0 and H_1, with selected critical region.

Thus it follows that H_1 should be accepted here because the sample value $\bar{x} = 194$ is to the right of $\bar{x}_0 = 193.8$.

With this choice of $\bar{x}$ as the borderline value for making decisions, the value of β becomes the area under the H_1 curve to the left of $\bar{x} = 193.8$. By the methods explained before, the value of β will be found to be .17. Figure 2 displays these results geometrically. Although the value of β is considerably larger than the value of α here, as contrasted to using the halfway point which made $\beta = \alpha = .10$, the archaeologist may consider the relative sizes of α and β to be satisfactory because he was much more concerned about making a type I error than about making a type II error. If the archaeologist should feel that the value of β is too large in relation to the value of α, all he would need to do is to decrease the value of $\bar{x}_0$ until he obtained a pair of values that were satisfactory to him in a relative sense.

The part of the $\bar{x}$ axis to the right of $\bar{x}_0$ is called the *critical region* of the test. It consists of those values of $\bar{x}$ that correspond to the rejection of H_0. The method for testing the hypothesis H_0 by means of $\bar{x}$ can be expressed very simply in terms of its critical region by stating that the hypothesis H_0 will be rejected if the sample value of $\bar{x}$ falls in the critical region of the test; otherwise H_0 will be accepted.

This method for testing the hypothesis H_0 is the method that will be used in this book for testing various hypotheses. It consists essentially of selecting a critical region for the variable being used to test the hypothesis such that the probability of the variable falling in the critical region is a fixed value α, and then agreeing to reject the hypothesis if, and only if, the sample value of the variable falls in the critical region. The experimental value of $\bar{x}$ is used only to make a decision after the critical region has been selected and is never permitted to influence the selection of the critical region.

In the foregoing problem α had the value .10 when the critical region was $\bar{x} > 193$ and the value .05 when the critical region was $\bar{x} > 193.8$. For problems of this type, the proper procedure is to choose the critical region so that the relative sizes of α and β are satisfactory; however, in many of the problems to come, this procedure would require lengthy computations and discussions of the relative importance of the two types of error involved. In order to avoid such lengthy discussions, a uniform procedure will be adopted of always choosing a critical region for which the value of α is .05. The value of $\alpha = .05$ is quite arbitrary here and some other value could have been agreed upon; however, it is the value of α most commonly used by applied statisticians. In any applied problem one can calculate the value of β and then adjust the value of α if the value of β is unsatisfactory when $\alpha = .05$. This works both ways, of course. For a very large experiment, with α fixed at .05, it might turn out that β is considerably smaller than .05. If the type I error were considered more serious than a type II error, then one would need to adjust the test to make α smaller than β, which would, of course, then make α smaller than .05.

This method of testing hypotheses requires one to choose a critical region for which $\alpha = .05$, but otherwise it does not determine how the critical region is to be chosen. Since it is desirable to make the probabilities of the two types of error as small as possible and since α is being fixed, one should choose a critical region that makes β as small as possible. Although the choice of the critical region in Fig. 2 was based on good sense, with the restriction that $\alpha = .05$, it can be shown that no other critical region with $\alpha = .05$ will have as small a value of β as the value $\beta = .17$.

Fortunately, in most simple problems an individual's good sense, or intuition, will lead him to a choice of a critical region that is the best possible in the sense that it will minimize the value of β. For more difficult problems in testing hypotheses, there is a mathematical theory that enables statisticians to find best critical regions. The critical regions that have been chosen in the problems to be solved in this and later chapters are the ones obtained by using this theory whenever it applies.

2.   TESTING A MEAN

The problem discussed in section 1 is an illustration of the general problem of testing the hypothesis that the mean of a particular normal population has a certain value. That problem was rather unusual in that there was only one alternative value for the mean. In most practical problems one has no specific information about the possible alternative values of the mean in case the value being tested is not the true value. The commonest situation is one in which all other values are possible. For such problems, the formulation corresponding to (1) assumes the form

(2)
$$H_0 : \mu = \mu_0$$
$$H_1 : \mu \neq \mu_0.$$

Here μ_0 denotes the particular value being tested. There are many practical problems, however, in which one is quite certain that if the mean is not equal to the value postulated under H_0 then its value must be larger than the postulated value. For such problems (2) would be replaced by

(3)
$$H_0 : \mu = \mu_0$$
$$H_1 : \mu > \mu_0.$$

For problems in which one is quite certain that if the mean is not equal to μ_0 then its value must be smaller than μ_0, one would, of course, replace $\mu > \mu_0$ by $\mu < \mu_0$ in (3).

As an illustration, suppose that a city has been purchasing brand A light bulbs for several years but is contemplating switching to brand B because of a better price. Salesmen for brand B claim that their product is just as good as brand A. Experience over several years has shown that brand A bulbs have a mean life of 1180 hours, with a standard deviation of 90 hours. To test the claim of the salesmen for brand B, 100 of their bulbs, purchased from regular retail sources, were tested. This sample yielded the values $\bar{x} = 1140$ and $s = 80$. Since mean burning time is a good measure of quality, the problem now is to test the hypothesis that the mean of brand B is equal to the brand A value against the alternative hypothesis that it has a smaller value. If the mean of brand B is denoted by μ, this test will assume the form of (3) with the inequality reversed, namely,

$$H_0 : \mu = 1180$$
$$H_1 : \mu < 1180.$$

This alternative was chosen because it was felt that if the quality of brand B bulbs were not the same as that of brand A bulbs then the brand B quality would undoubtedly be lower than the brand A quality. Salesmen are not likely to underrate their own products. Therefore, if these salesmen are telling the truth, H_0 will be true. If they are not telling the truth, their brand will be of lower quality because no salesman would be so stupid as to claim only equality when he could actually claim superiority for his product.

Now, good sense would suggest that the further $\bar{x}$ is to the left of the postulated mean of 1180 the less faith one should have in the truth of H_0 and the more faith one should have in some smaller value of μ being the true mean. Thus it is clear that the critical region should consist of small values of $\bar{x}$ and therefore of that part of the $\bar{x}$ axis to the left of some point $\bar{x}_0$. The problem therefore is to determine the point $\bar{x}_0$ so that the value of α will be .05. The technique for doing this is the same as that employed in solving the archaeologists' problem. Since $n = 100$ here, it follows that

$$\sigma_{\bar{x}} = \frac{\sigma}{\sqrt{100}} = \frac{\sigma}{10}.$$

If the hypothesis being tested here were the hypothesis of equal quality with respect to variability as well as mean burning time, rather than merely equal mean burning time, then it would be proper to use $\sigma = 90$ in this expression for $\sigma_{\bar{x}}$. However, since only the means are being assumed equal, the sample estimate $s = 80$ is used instead; consequently

$$\sigma_{\bar{x}} \doteq \frac{80}{10} = 8.$$

Since the 5 per cent left tail area of a standard normal curve lies to the left of the point $z = -1.64$, it follows that $\bar{x}_0$ is a point 1.64 standard deviations to the left of the mean $\mu = 1180$. The standard deviation here is $\sigma_{\bar{x}} \doteq 8$; hence the desired critical region is that part of the $\bar{x}$ axis to the left of

$$1180 - 1.64(8) \doteq 1167.$$

These results are displayed in Fig. 3. If you prefer to use algebra to obtain the value of $\bar{x}_0$, you should proceed as in the archaeologists' problem and write down the equation

$$-1.64 = \frac{\bar{x} - 1180}{8}.$$

Solving for $\bar{x}$ will, of course, yield the solution $\bar{x}_0 = 1167$.

Now that the critical region has been selected, one can proceed to test the hypothesis H_0. Since the sample value $\bar{x} = 1140$ falls in the critical region, the hypothesis H_0 will be rejected. It seems quite certain that a sample mean as low as 1140 could not have been obtained from a random sample of size 100 taken from a population with mean 1180. This implies that the salesmen of brand B bulbs are not justified in their claim of the same quality as brand A. Since it

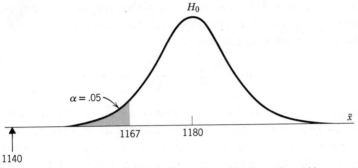

FIGURE 3.   Critical region for testing H_0.

is quite certain that μ is less than 1180, one should consider next the question of how much less. If a point estimate of μ were desired, then, of course, $\bar{x} = 1140$ would be selected as the estimate. One could also find a confidence interval for μ and then determine the maximum and minimum differences that are likely to exist between the two population means. Such considerations would be necessary before one could decide whether the lower price for brand B would compensate for the lower quality. Since the object of this section is to explain how to test hypotheses, these practical matters are not discussed here; however, the solutions of actual problems by statistical methods usually require such considerations.

   As an illustration for which formulation (2) would be preferred to (3), consider the following problem. Records for the last several years of freshmen admitted to a certain college showed that their mean score on an aptitude test was 115. An administrator who is interested in knowing whether the new freshman class is a typical class with respect to aptitude proposes to test the hypothesis that the new freshman class mean is the same as that of former classes. Since he has no reason for believing that the new class is any better or any worse than former classes, he should use formulation (2). This becomes

$$H_0: \mu = 115$$
$$H_1: \mu \neq 115.$$

For the purpose of testing this hypothesis, the aptitude test score of every tenth student is obtained from the admissions office. Suppose this yielded a sample of size $n = 50$ and that for this sample the mean and standard deviation turned out to be $\bar{x} = 118$ and $s = 20$.

   Since the further $\bar{x}$ is from the hypothetical mean value of 115, whether to the right or the left, the less faith one would have in the truth of H_0, it is clear that the critical region here should consist of values of $\bar{x}$ out in the two tails of the $\bar{x}$ curve centered at 115. Now, for a sample as large as 50, $\bar{x}$ may be assumed to be normally distributed. Furthermore, since $s = 20$, the standard deviation of $\bar{x}$ may be approximated as follows:

$$\sigma_{\bar{x}} = \frac{\sigma}{\sqrt{50}} \doteq \frac{20}{\sqrt{50}} \doteq 2.8.$$

Since the probability is .05 that $\bar{x}$ will assume a value more than 1.96 standard deviations away from the mean, it follows that the desired critical region of size $\alpha = .05$ should consist of the values of $\bar{x}$ out in the two tails of the $\bar{x}$ curve determined by the two values $115 - 1.96(2.8) = 109.5$ and $115 + 1.96(2.8) = 120.5$. These results are displayed in Fig. 4.

Since $\bar{x} = 118$ yields a point, indicated on Fig. 4 by an arrow, that does not fall in the critical region, the hypothesis H_0 will be accepted. The college administration may relax in the knowledge that the new freshman class is at about the same level of aptitude as former classes.

The acceptance of a hypothesis in this manner is a practical decision matter. It does not imply that one believes that the hypothesis is precisely correct, and it certainly is not a proof of the truth of the hypothesis. Rather, it implies that the sample data are compatible with the postulated value of the mean. From a practical point of view, it makes little difference whether the true mean has the postulated value or whether it has a value close to the postulated value. How close the true value of the mean must be to the postulated value in order that the hypothesis be accepted can be determined by the confidence interval methods explained in Chapter 6. In view of these remarks, accepting a hypothesis is to be construed as admitting that the hypothesis is reasonably close to the true situation and that, from a practical point of view, one may therefore treat it as representing the true situation.

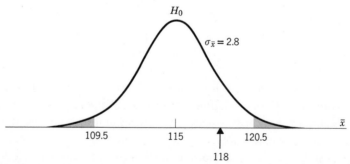

FIGURE 4.   Two-sided critical region of size $\alpha = .05$.

After the administrator notices that the sample mean is higher than the old mean, he will undoubtedly wish to claim that the new class is better than the former classes. In view of this fact, there would be the temptation to treat this problem as one of testing $H_0: \mu = 115$ against $H_1: \mu > 115$ and to use a one-sided critical region as in the earlier problems; however, this would be illegal because the decision as to the possible alternative values must be based on knowledge other than that given by the sample. A simple way of deciding whether to use a one-sided or two-sided test is to ask oneself what the alternative values of interest are before the sample has been taken or, what is equivalent, before the sample results have been observed.

The sizes of the type II errors in these problems have not been calculated because that would require considerably more calculation and discussion of the problems.

A useful application of the idea of testing a mean arises in industrial quality control work. Suppose that a machine is turning out a large number of parts that are used in some manufactured article and that it is important for the diameter of such a part to be very accurate. It is customary for an inspector to sample periodically from the production line to see whether the diameters are behaving properly. If 5 parts are measured every hour and their sample mean recorded, a large number of $\bar{x}$ values will be obtained after a few weeks of inspection. The mean of all these $\bar{x}$'s may be treated as the true mean of the population of diameters of parts and the standard deviation of these $\bar{x}$'s as the true value of $\sigma_{\bar{x}}$.

By treating the preceding values as true values, one can calculate the values $\mu - 3\sigma_{\bar{x}}$ and $\mu + 3\sigma_{\bar{x}}$. From normal curve properties, the probability is .997 that a sample value of $\bar{x}$ will fall between these two limits; therefore, if an $\bar{x}$ value falls outside this interval, there is good reason to believe that something has gone wrong with the machine turning out the parts. Experience has shown that a machine that is operating properly will behave very much like a random number machine in the sense that successive parts turned out behave very much like random samples from a population of parts. A three-standard deviation interval is used instead of, say, a two-standard deviation interval because about 5 per cent of the $\bar{x}$ values would fall outside a two-standard deviation interval, even though the machine is operating satisfactorily, and therefore the inspector would be looking for trouble too often when there is none. Furthermore, an industrial machine is only an approximation of an ideal random number machine. Experience indicates that a three-standard deviation interval is about right from a practical point of view.

After data have been gathered for a few weeks so that the limits $\mu - 3\sigma_{\bar{x}}$ and

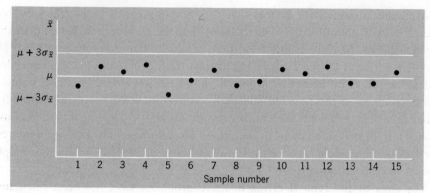

FIGURE 5.   Control chart for the mean.

$\mu + 3\sigma_{\bar{x}}$ can be obtained, the control chart is ready to be constructed. It merely consists of a horizontal band with the horizontal axis marked off with sample numbers. A control chart of this type is shown in Fig. 5.

Each point corresponds to a sample value of $\bar{x}$ obtained from the five parts selected each hour, after the initial data gathering period. It will be observed that the process appears to be under control. The striking advantage of a control chart is that it warns the inspector by means of probability of trouble with a machine before it has turned out a large number of bad parts, which otherwise might not be discovered until some time later when the parts were being used in assembling the article being manufactured.

2.1. Small Sample Method

In the problem of testing the mean aptitude score of freshmen, discussed in section 2, the sample size was large enough to justify the approximation used there. If the sample size had been considerably smaller than 50, the replacement of σ by s might have introduced a serious error. For such small sample problems, one can use Student's t distribution in the same manner that normal z is used for large samples. For example, suppose n had been equal to 20 in the aptitude problem. Then one would have calculated the value of t given by formula (4) in Chapter 7. For this problem,

$$t = \frac{\bar{x} - \mu}{s} \sqrt{n} = \frac{118 - 115}{20} \sqrt{20} = .67.$$

From Table V in the appendix, it will be found that for $\nu = 19$ degrees of freedom the .025 value of t is 2.093. Since a two-sided critical region is being used here, it follows that it should consist of those values of t numerically larger than 2.093. Because $t = .67$ does not fall in the critical region, H_0 is accepted. Since the derivation of Student's t distribution requires the assumption that the basic variable x is normally distributed, one must be a little careful when applying it to small samples to make certain that x possesses an approximate normal distribution. The large sample method does not require this precaution, because $\bar{x}$ is likely to be very nearly normally distributed, even for moderate size samples; however, the large sample method has the serious fault of requiring a knowledge of σ or a good estimate of it, and this is not likely to be available in small sample practical problems.

3.	TESTING THE DIFFERENCE OF TWO MEANS

The problem of the light bulbs that was solved in section 2 can be modified slightly to produce a problem that is typical of many in real life. Suppose the city buying light bulbs had no experience with either brand A or brand B bulbs and wished to decide which brand to purchase, the prices being the same. It would then be necessary to test a sample of each brand, rather than just a sample of brand B as in the earlier problem. Suppose a sample of 100 bulbs from each of the two brands is tested and that the samples yield the values $\bar{x}_1 = 1160$, $s_1 = 90$, $\bar{x}_2 = 1140$, and $s_2 = 80$, in which the subscripts 1 and 2 refer to brands A and B, respectively.

Now since brand A yields a larger mean burning time than brand B, it would appear that brand A is superior to brand B; however, it might be that the reverse is true, but some bad luck with a few of the bulbs of brand B produced an unusually low sample mean. A second set of samples of 100 each might conceivably produce different results. The problem therefore reduces to determining whether this difference of sample means, namely $\bar{x}_1 - \bar{x}_2$, is large enough to justify the belief that brand A is superior to brand B.

In order to solve this problem, it is necessary to know how $\bar{x}_1 - \bar{x}_2$ varies if repeated sampling experiments of the same kind are performed. Each sampling experiment consists of taking a sample of 100 bulbs from brands A and B, testing the 200 bulbs, determining the values of $\bar{x}_1$ and $\bar{x}_2$, and recording the value of the variable $\bar{x}_1 - \bar{x}_2$. If a large number of such sampling experiments were carried out, a large number of values of the variable $\bar{x}_1 - \bar{x}_2$ would be obtained. These values could be classified into a frequency table, and a histogram drawn, to give one a good idea of the limiting, or theoretical, distribution of $\bar{x}_1 - \bar{x}_2$. As in the case of a single mean, $\bar{x}$, it is not necessary to carry out these sampling experiments because the form of the limiting distribution can be worked out mathematically. It can be shown that $\bar{x}_1 - \bar{x}_2$ will possess a normal distribution if x_1 and x_2 do, with a mean equal to the difference of the population means of x_1 and x_2, namely $\mu_1 - \mu_2$, and with a variance (square of standard deviation) equal to the sum of the variances of $\bar{x}_1$ and $\bar{x}_2$. This result is expressed in the form of a theorem:

(4)   **Theorem.**   *If x_1 and x_2 possess independent normal distributions with means μ_1 and μ_2 and standard deviations σ_1 and σ_2, then the variable $\bar{x}_1 - \bar{x}_2$ will possess a normal distribution with mean $\mu_1 - \mu_2$ and standard deviation given by the formula*

$$\sigma_{\bar{x}_1 - \bar{x}_2} = \sqrt{\sigma_{\bar{x}_1}^2 + \sigma_{\bar{x}_2}^2} = \sqrt{\frac{\sigma_1^2}{n_1} + \frac{\sigma_2^2}{n_2}}.$$

Here n_1 and n_2 are the sample sizes on which $\bar{x}_1$ and $\bar{x}_2$ are based. It should be noted that the theorem requires that the two populations being sampled be normal; however, if n_1 and n_2 are as large as 25, say, then $\bar{x}_1$ and $\bar{x}_2$ may be assumed to be normally distributed, in which case $\bar{x}_1 - \bar{x}_2$ will also be normally distributed. It should also be noted that x_1 and x_2 are required to be independent variables. For example, if the problem had been to test the hypothesis that there is no difference between the mean length of people's right feet and left feet, it would not be correct to let x_1 represent the length of an individual's right foot and x_2 the length of his left foot because a large value of x_1 would certainly increase the probability of a large value of x_2. If, however, the x_2 values were obtained from a different set of individuals, rather than those for x_1, then x_1 and x_2 would be independent variables. The proper way to treat paired data, such as would arise in measuring both feet of each individual, would be to take the difference of each pair and then use the earlier methods for testing whether the mean of the variable $x_1 - x_2$ is zero. The test would then be based on a sample of size $n = n_1 = n_2$.

Problems such as the selection of light bulbs can be solved by means of the theorem in (4), provided they are treated as problems of testing the appropriate hypothesis. From a practical point of view, it should make little difference whether one rejects the hypothesis that the brands are equally good or accepts the hypothesis that they differ in quality. From a theoretical point of view, however, it is more convenient to test the hypothesis that the brands are equally good than the hypothesis that they differ in quality. As a result, one sets up the hypothesis

(5) $$H_0: \mu_1 - \mu_2 = 0.$$

An equivalent way of writing this is

$$H_0 : \mu_1 = \mu_2.$$

This type of hypothesis is known as a *null* hypothesis because it assumes that there is no difference. Very often, however, the experimenter believes that there is an appreciable difference and hopes that the sample evidence will reject the hypothesis. If the sample does reject the hypothesis, then one can claim with justification that a real difference in population means exists. If the sample does not reject the hypothesis, then there is a fair probability that the sample difference is caused by sampling variation, under the assumption that the population means are equal.

In the light of the theorem in (4) and the preceding discussion, testing the hypothesis given by (5) is equivalent to testing the hypothesis that the mean of the normal variable $\bar{x}_1 - \bar{x}_2$ is 0. But this type of problem was solved in section 2. Since the alternative hypothesis would ordinarily be chosen as

$$H_1 : \mu_1 \neq \mu_2,$$

it follows that the same methods should be applied here as in the solution of the aptitude test problem which used the formulation in (2). Now, from the theory in (4) the variable $\bar{x}_1 - \bar{x}_2$ may be assumed to be normally distributed with mean 0, because $\mu_1 = \mu_2$ under H_0, and with standard deviation given by

$$\sigma_{\bar{x}_1 - \bar{x}_2} = \sqrt{\frac{\sigma_1^2}{n_1} + \frac{\sigma_2^2}{n_2}}.$$

Unfortunately, the population values σ_1^2 and σ_2^2 are unknown; consequently, they must be approximated by their sample estimates, namely $s_1^2 = (90)^2$ and $s_2^2 = (80)^2$. Since $n_1 = n_2 = 100$ here, the approximate value of the standard deviation becomes

$$\sigma_{\bar{x}_1 - \bar{x}_2} \doteq \sqrt{\frac{8100}{100} + \frac{6400}{100}} \doteq 12.$$

A critical region based on equal tail areas under the normal curve for $\bar{x}_1 - \bar{x}_2$ is chosen for which $\alpha = .05$. This means that the critical region consists of that part of the horizontal axis lying more than $1.96\sigma_{\bar{x}_1 - \bar{x}_2} = 24$ units away from 0. Figure 6 shows geometrically the distribution of $\bar{x}_1 - \bar{x}_2$ and the selected critical region.

Since the two samples of 100 each yielded the value $\bar{x}_1 - \bar{x}_2 = 1160 - 1140 = 20$ and since 20 does not fall in the critical region for this test, the hypothesis is accepted.

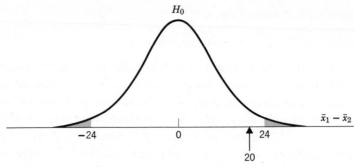

FIGURE 6. Distribution of $\bar{x}_1 - \bar{x}_2$ when $\mu_1 = \mu_2$.

Just as in the earlier problem of testing a single mean, the acceptance of a hypothesis in this manner does not imply that one believes that the hypothesis is true. It does imply, however, that one is not convinced by the sample evidence that there is an appreciable difference and that, unless further evidence is presented to the contrary, one is willing to assume that for all practical purposes there is no appreciable difference in the population means. It is a mathematical convenience to formulate a hypothesis in this manner. It would be more realistic to test whether the means differed by less than a specified amount, but the resulting theory would be much more complicated. A student should not deceive himself into believing that he has proved the hypothesis to be true just because he has agreed to accept it.

In view of the fact that sample estimates were needed as approximations for population variances, the methods used here are large sample methods.

There are several words and phrases used in connection with testing hypotheses that should be brought to the attention of students. When a test of a hypothesis produces a sample value falling in the critical region of the test, the result is said to be *significant*; otherwise one says that the result is *not significant*. This word arises from the fact that such a sample value is not compatible with the hypothesis and therefore signifies that some other hypothesis is necessary. The probability of committing a type I error, which is denoted by α, is called the *significance level* of the test. For problems being solved routinely in this book, the significance level has been chosen equal to .05.

If one analyzes the technique that has been used to test the various hypotheses that have been treated thus far he will observe that it is merely a rule for making a decision. This rule is usually based on the sample value of some random variable and consists in dividing all the possible values of the random variable into two groups, those associated with the rejection of H_0 and which form what is called the critical region of the test, and those associated with the acceptance of H_0.

From this general point of view, a test of a hypothesis is merely a systematic way of making a practical decision and there is no implication made concerning the truth or falsity of the hypothesis being treated.

▶3.1. Small Sample Method

If the sample sizes are too small to justify replacing σ_1 and σ_2 by their sample estimates in the preceding test, then the appropriate Student t test may be used. For testing the difference of two means, the theory of Student's t distribution requires one to assume that the two basic variables x_1 and x_2 possess independent normal distributions with equal standard deviations. These assumptions are considerably more restrictive than those needed for the large sample method. If these assumptions are reasonably satisfied, then one may treat the variable

$$t = \frac{\bar{x}_1 - \bar{x}_2 - (\mu_1 - \mu_2)}{\sqrt{(n_1 - 1)s_1^2 + (n_2 - 1)s_2^2}} \sqrt{\frac{n_1 n_2 (n_1 + n_2 - 2)}{n_1 + n_2}}$$

as a Student t variable with $\nu = n_1 + n_2 - 2$ degrees of freedom. The solution is now carried out in the same manner as for testing a single mean. For example, if the problem just solved is altered to make the sample sizes 10 each, then the value of t will become

$$t = \frac{1160 - 1140}{\sqrt{9(90)^2 + 9(80)^2}} \sqrt{\frac{100(18)}{20}} = .53.$$

From Table V in the appendix it will be found that the 5 per cent critical value of t corresponding to $\nu = 18$ degrees of freedom is 2.10. Since the value $t = .53$ falls inside the noncritical interval, which extends from -2.10 to $+2.10$, the hypothesis is accepted. Because the hypothesis was accepted before for a much larger sample, it would obviously be accepted here as well.

Modifications of the foregoing t test exist for problems in which it is unreasonable to assume that the two variances are equal; however, they are not considered here.

4. TESTING A PROPORTION

The large sample normal curve methods employed to solve estimation problems for binomial p can be employed also to test hypotheses about p. As a result, the techniques for testing the hypothesis that p has a fixed value or that two proportions are equal are much the same as those explained in the last two sections for means. As an illustration, consider the following genetics problem. According to Mendelian inheritance theory, certain crosses of peas should give yellow and green peas in a ratio of 3:1. In an experiment 176 yellow and 48 green peas were obtained. Are these numbers compatible with Mendelian theory?

This problem may be considered as a problem of testing the hypothesis

$$H_0 : p = \frac{3}{4}$$

in which p denotes the probability that a pea selected at random will be yellow. The 224 peas may be treated as 224 trials of an experiment for which $p = \frac{3}{4}$ is the probability of success in a single trial. From formula (7), Chapter 5, it follows that $\widehat{p} = x/n$ may be treated as a normal variable with mean $p = \frac{3}{4}$ and standard deviation given by

$$\sigma_{\widehat{p}} = \sqrt{\frac{pq}{n}} = \sqrt{\frac{\frac{3}{4} \cdot \frac{1}{4}}{224}} = .029.$$

The problem now is much the same as the problem of testing a normal mean. The critical region here is chosen as those values of $\widehat{p}$ in the two tails of the normal curve for $\widehat{p}$. For $\alpha = .05$, the critical region will then consist of those values of $\widehat{p}$ lying outside the interval given by

$$p - 1.96 \sqrt{\frac{pq}{n}} \quad \text{and} \quad p + 1.96 \sqrt{\frac{pq}{n}}.$$

Since $n = 224$ and $p = \frac{3}{4}$ here, computations will yield the interval (.693, .807). Figure 7 shows the approximate normal distribution for $\widehat{p}$ and the critical region just determined. Since the sample value $\widehat{p} = 176/224 = .79$ does not fall in the critical region, the hypothesis H_0 is accepted. Thus, on the basis of these data, there is no reason for doubting that Mendelian inheritance is operating here.

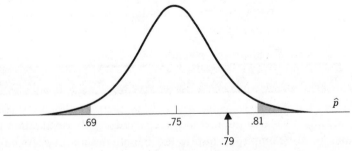

FIGURE 7.   Approximate normal distribution of $\hat{p}$.

As a second illustration of how to use normal curve methods to test a hypothesis about binomial p, consider the problem discussed briefly near the end of Chapter 5. A politician had claimed a 60 per cent backing on a piece of legislation and a sample of 400 voters had been taken to check this claim. The question then arose as to how small the sample percentage would need to be before the claim could be rightfully refuted. This problem can be considered as a problem of testing the hypothesis

$$H_0:p = .6.$$

Since the interest in this problem centers on whether $p = .6$, as against the possibility that $p < .6$, this problem is somewhat like that of the light bulbs discussed in the first section of this chapter in that the alternatives are all on one side of the hypothetical value and therefore that the critical region should be under one tail of the proper normal curve. The natural alternative hypothesis here is

$$H_1:p < .6.$$

The critical region of size $\alpha = .05$ for this problem should therefore be selected to be under the left 5 per cent tail of the normal curve whose mean is $p = .6$ and whose standard deviation is given by

$$\sigma_{\hat{p}} = \sqrt{\frac{pq}{n}} = \sqrt{\frac{(.6)(.4)}{400}} = .0245.$$

Since, from Table IV, 5 per cent of the area of a standard normal curve lies to the left of $z = -1.64$, this means that the critical region should consist of all those values of $\hat{p}$ that are smaller than the value of $\hat{p}$ which is 1.64 standard deviations to the left of the mean. For this problem, the critical region therefore consists of all those values of $\hat{p}$ that are smaller than

$$p - 1.64\sqrt{\frac{pq}{n}} = .6 - 1.64(.0245) = .56.$$

If the sample value of $\hat{p}$ turned out to be less than .56, the politician's claim would be rejected.

Another useful application of testing a binomial p arises in industrial control charts for the percentage of defective parts in mass production of parts. This application is not confined to industrial problems; it may be used wherever one has repeated operations. The technique is precisely the same as for control charts for the mean. One uses accumulated experience to obtain a good estimate of p;

then one constructs the control band given by $p - 3\sqrt{pq/n}$ and $p + 3\sqrt{pq/n}$. Here n is the size sample on which each plotted proportion is based.

As an illustration of how one would construct such a chart, consider the following problem. A record is kept for ten days of the number of words mistyped by students learning typing. During that period of time they typed a total of approximately 20,000 words, of which 800 were mistyped. The problem is to use these data to construct a control chart for the proportion of errors made per class hour by a student who types approximately 600 words per class hour. Assuming the given student is typical, a good estimate of p is given by dividing the total number of mistyped words by the total number of typed words. This estimate is

$$p = \frac{800}{20{,}000} = .04.$$

Since the proportions to be plotted on the control chart are those for an hour's typing, it follows that $n = 600$ here. If these values are substituted in the formulas given in the preceding paragraph, the desired lower and upper boundaries for the control chart will become

$$.04 - 3\sqrt{\frac{(.04)(.96)}{600}} \quad \text{and} \quad .04 + 3\sqrt{\frac{(.04)(.96)}{600}}.$$

These simplify to .016 and .064. The chart can now be constructed in the same manner as for the mean, except that now one plots the proportion of mistyped words every hour rather than the sample mean.

5. TESTING THE DIFFERENCE OF TWO PROPORTIONS

A problem of much importance and frequent occurrence in statistical work is the problem of determining whether two populations differ with respect to a certain attribute. For example, is there any difference in the percentages of smokers and nonsmokers among those who have heart ailments?

Problems of this type can be treated as problems of testing the hypothesis

$$H_0 : p_1 = p_2,$$

in which p_1 and p_2 are the two population proportions of the attribute. If n_1 and n_2 denote the size samples taken and $\widehat{p}_1$ and $\widehat{p}_2$ the resulting sample proportions obtained, then the variable to use in solving this problem is $\widehat{p}_1 - \widehat{p}_2$. This corresponds to using $\bar{x}_1 - \bar{x}_2$ in the problem of testing the hypothesis that $\mu_1 = \mu_2$.

The methods used to solve that problem can be employed here as well because $\hat{p}_1 - \hat{p}_2$ may be considered as being approximately normally distributed with mean $p_1 - p_2$ and with standard deviation given by

$$\sigma_{\hat{p}_1-\hat{p}_2} = \sqrt{\frac{p_1 q_1}{n_1} + \frac{p_2 q_2}{n_2}}.$$

When testing the hypothesis $H_0 : p_1 = p_2$, the mean of the distribution of $\hat{p}_1 - \hat{p}_2$ will, of course, be equal to 0.

As an illustration of how to use these formulas, consider the following problem. A sample of 400 sailors was split into two equal groups by random selection. One group was given brand A pills of a seasickness preventive, and the other brand B pills. The number in each group that refrained from becoming seasick during a heavy storm was 152 and 132. Can one conclude that there is no real difference in the effectiveness of these pills?

Calculations give

$$\hat{p}_1 = \frac{152}{200} = .76, \qquad \hat{p}_2 = \frac{132}{200} = .66,$$

$$\sigma_{\hat{p}_1-\hat{p}_2} = \sqrt{\frac{p_1 q_1}{200} + \frac{p_2 q_2}{200}}.$$

Since the values of p_1 and p_2 are unknown, they must be approximated by sample estimates. Although the values are unknown, they are assumed to be equal under the hypothesis $H_0 : p_1 = p_2$. If this common value is denoted by p, then a good estimate of p is the value obtained from the sample proportion of the combined data. There were 284 of 400 sailors who were successes in this total experiment; hence p would be estimated by means of ~ *estimation of p*

$$\hat{p} = \frac{284}{400} = .71.$$

By replacing p_1 and p_2 by p in the formula for the standard deviation and then approximating p by $\hat{p}$, one obtains

$$\sigma_{\hat{p}_1-\hat{p}_2} = \sqrt{(.71)(.29)(\tfrac{1}{200} + \tfrac{1}{200})} = .045.$$

The use of a two-sided critical region of size .05 yields the critical region displayed in Fig. 8. Since $\hat{p}_1 - \hat{p}_2 = .10$ here, it falls in the critical region and therefore the hypothesis is rejected. It would appear that brand A gives somewhat better protection against seasickness than brand B, at least for sailors in stormy weather.

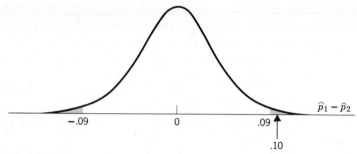

FIGURE 8. Distribution and critical region for $\hat{p}_1 - \hat{p}_2$.

6.	ADDITIONAL ILLUSTRATIONS

1.   Assume that x possesses a normal distribution with $\sigma = 6$ but that its mean is unknown. If a sample of size 16 is taken and $\bar{x} = 33$, (a) 2 test the hypothesis $H_0 : \mu = 30$ against $H_1 : \mu > 30$. (b) 3 A second sample of size 32 was taken some time later and yielded a mean of 29; hence there is reason to believe that the mean has changed. Test the hypothesis that the mean has not changed against the possibility of its having slipped to the left. The solutions follow.

(a) $$\sigma_{\bar{x}} = \frac{6}{\sqrt{16}} = 1.5, \quad z = \frac{33 - 30}{1.5} = 2.00.$$

Since large values of $\bar{x}$ would favor H_1, the critical region should consist of large values of $\bar{x}$, and hence large values of z. For $\alpha = .05$ the critical region is therefore given by $z > 1.64$. Since $z = 2.00$ lies in the critical region, reject H_0.

(b) $$\sigma_{\bar{x}_1 - \bar{x}_2} = \sqrt{36\left[\frac{1}{16} + \frac{1}{32}\right]} = 1.84$$

$$z = \frac{29 - 33}{1.84} = -2.17.$$

Here the left tail of the distribution of $\bar{x}_1 - \bar{x}_2$ should be chosen as the critical region because it favors slippage to the left. For $\alpha = .05$ this corresponds to choosing $z < -1.64$ as the critical region. Since $z = -2.17$ lies in the critical region, reject H_0.

2.   The following values were obtained by sampling a normal population with $\mu = 3.0$ and $\sigma = .5$. Twenty-five samples each were taken by two different sam-

plers, A and B. Using these data solve the following problems. (a) 2 Combine the data into one set and test the hypothesis $H_0:\mu = 3.0$ against $H_1:\mu \neq 3.0$. (b) 2 Construct a control chart for these 50 observational values. (c) 3 Test the hypothesis that $\mu_A = \mu_B$ using the fact that $\sigma_A = \sigma_B = \sigma = .5$. (d) 3 Work part (c) without assuming that the standard deviations are known. (e) 2.1 Using the data from A and the t distribution, test the hypothesis $H_0:\mu_A = 3.3$ against $H_1:\mu < 3.3$. (f) 3.1 Using the t distribution test the hypothesis $H_0:\mu_A = \mu_B$ and compare your t value with the z value of part (d).

A	3.5	3.4	2.8	2.6	2.3	2.4	2.9	2.9	3.1	3.3	3.9	3.0	2.9	3.0	2.5
B	2.9	3.6	3.1	2.5	2.6	3.6	2.9	2.1	2.6	2.5	2.3	2.6	3.3	3.9	2.7

A	2.7	3.9	2.6	2.8	3.2	3.6	2.7	3.0	3.8	3.1
B	2.4	2.8	3.0	3.3	3.7	3.3	3.!	2.7	2.4	3.3

(a) Calculations for all 50 values give $\bar{x} = 2.98$; hence

$$z = \frac{\bar{x} - \mu}{\sigma_{\bar{x}}} = \frac{2.98 - 3.00}{\dfrac{.5}{\sqrt{50}}} = -.28.$$

Since $|z| < 1.96$, H_0 is accepted.

(b) A three-standard deviation band will have the limits $\mu \pm 3\sigma = 3.0 \pm 1.5 = (1.5, 4.5)$. This band includes all 50 points, since no values are smaller than 1.5 and none larger than 4.5.

(c) Calculations give $\bar{x}_A = 3.04$ and $\bar{x}_B = 2.93$; hence

$$z = \frac{\bar{x}_A - \bar{x}_B}{\sqrt{\sigma^2\left(\dfrac{1}{25} + \dfrac{1}{25}\right)}} = \frac{.11}{\sqrt{.25\left(\dfrac{1}{25} + \dfrac{1}{25}\right)}} = .78.$$

Since $|z| < 1.96$, H_0 is accepted.

(d) Additional calculations give $s_A^2 = .202$ and $s_B^2 = .205$; hence

$$z = \frac{\bar{x}_A - \bar{x}_B}{\sqrt{\dfrac{s_A^2}{25} + \dfrac{s_B^2}{25}}} = \frac{-.11}{\sqrt{\dfrac{.202}{25} + \dfrac{.205}{25}}} = .86.$$

Since $|z| < 1.96$, H_0 is accepted.

(e) Here $\bar{x} = 3.04$, $s^2 = .202$, and $\nu = 24$; hence

$$t = \frac{3.04 - 3.30}{\sqrt{.202}}\sqrt{25} = -2.89.$$

Since $t < -1.7109$, reject H_0. This is the correct decision because $\mu_A = 3.0$.

(f) Here

$$t = \frac{3.04 - 2.93}{\sqrt{4.86 + 4.91}}\sqrt{\frac{(25)(25)(48)}{25 + 25}} = .86.$$

Since $|t| < 2.0639$, accept H_0. The t and z values are the same to this degree of calculation accuracy. For samples as large as this there is very little difference between the two tests.

3.   A sample of size 200 car owners in a certain city showed that 48 of them had driver's licenses that had expired. Using these data solve the following problems. (a) 4 Test the hypothesis that the population proportion is $p = .30$. (b) 4 Find the boundaries for a control chart for $\hat{p}$ if daily samples of size 50 are to be taken and if the population proportion is $p = .30$.

(a) $\hat{p} = \dfrac{48}{200} = .24$; hence

$$z = \frac{.24 - .30}{\sqrt{\dfrac{(.30)(.70)}{200}}} = -1.85.$$

Since $|z| < 1.96$, accept H_0.

(b) Since $n = 50$, the boundaries are given by

$$p \pm 3\sqrt{pq/50} = .30 \pm 3\sqrt{\frac{(.30)(.70)}{50}} = 30 \pm .19;$$

hence they are .11 and .49.

4.   Referring to problem 3, suppose a sample of size 100 car owners was taken from another section of the city and showed that 33 of them had lapsed driver's licenses. (a) 5 Test the hypothesis $H_0 : p_1 = p_2$, where p_1 and p_2 refer to the first and second sample populations, respectively. (b) 5 Work part (a) if the alternative hypothesis is $H_1 : p_1 < p_2$.

(a) $\hat{p}_1 = .24$, $\hat{p}_2 = .33$, $\hat{p} = \dfrac{48 + 33}{200 + 100} = .27$; hence

$$z = \frac{.24 - .33}{\sqrt{(.27)(.73)\left[\dfrac{1}{200} + \dfrac{1}{100}\right]}} = -1.66.$$

Since $|z| < 1.96$, accept H_0.

(b) Here the critical region is given by $z < -1.64$. Since $z < -1.64$, reject H_0.

EXERCISES

Section 1

1. In a court case in which an individual is being tried for theft, what are the two types of error? Which type of error is considered by society more important?

2. Give an illustration of a hypothesis for which the type II error would be considered much more serious than the type I error.

3. Suppose you agree to reject a hypothesis if two tosses of an honest coin produce two heads. What are the sizes of the two types of error?

4. A coin is tossed twice. Let x denote the number of heads obtained. Consider the hypothesis $H_0:p = .5$ and the alternative $H_1:p = .7$, in which p is the probability of obtaining a head, and assume that one is going to test this hypothesis by means of the value of x. The distributions of x under H_0 and H_1 can be obtained by means of the binomial distribution formula (1) in Chapter 5. Suppose one chooses $x = 2$ as the critical region for the test. Calculate the sizes of the two types of error.

5. In problem 4 suppose $x = 0$ had been chosen as the critical region for the test. Now calculate the sizes of the two types of error and compare your results with those of problem 4. Comment about these two choices of critical region.

Section 2

6. Given $\bar{x} = 82$, $\sigma = 15$, and $n = 100$, test the hypothesis that $\mu = 86$.

7. Given $\bar{x} = 82$, $\sigma = 15$, and $n = 25$, test the hypothesis that $\mu = 86$.

8. A purchaser of bricks believes that the quality of the bricks is deteriorating. From past experience, the mean crushing strength of such bricks is 400 pounds, with a standard deviation of 20 pounds. A sample of 100 bricks yielded a mean of 395 pounds. Test the hypothesis that the mean quality has not changed against the alternative that it has deteriorated.

9. Many years of experience with a university entrance examination in English yielded a mean score of 64 with a standard deviation of 9. All the students from a certain city, of which there were 54, obtained a mean score of 68. Can one be quite certain that students from this city are superior in English?

10. A manufacturer of fishing line claims that his 5-pound test line will average 8 pounds test. Is he justified in his claim if a sample of size 50 yielded $\bar{x} = 7$ pounds and $s = 1.4$ pounds?

11. Construct a control chart for $\bar{x}$ for the following data on the blowing time of fuses, samples of 5 being taken every hour. Each set of 5 has been arranged in order of magnitude. Estimate μ by calculating the mean of all the data and estimate $\sigma_{\bar{x}}$ by first estimating σ by means of s calculated for all 60 values. State whether control seems to exist here.

42	42	19	36	42	51	60	18	15	69	64	61
65	45	24	54	51	74	60	20	30	109	91	78
75	68	80	69	57	75	72	27	39	113	93	94
78	72	81	77	59	78	95	42	62	118	109	109
87	90	81	84	78	132	138	60	84	153	112	136

12. Take a sample of size 25 from a table of one-digit random numbers (Table II in the appendix) and test the hypothesis that $\mu = 4.5$. Use the fact that $\sigma = 2.87$ for this distribution and that $\bar{x}$ may be treated as a normal variable. Bring your result to class. Approximately 95 per cent of the class should accept this hypothesis because it is true.

Section 2.1

13. Given that x is normally distributed and given the sample values $\bar{x} = 42$, $s = 6$, $n = 20$, test the hypothesis that $\mu = 44$, using the t distribution.

14. The following data give the corrosion effects in various soils for coated and uncoated steel pipe. Taking differences of pairs of values, test by means of the t distribution the hypothesis that the mean of such differences is zero.

Uncoated	42	37	61	74	55	57	44	55	37	70	52	55
Coated	39	43	43	52	52	59	40	45	47	62	40	27

15. Work problem 12, choosing $\alpha = .20$, using a sample of size 10 and assuming that the value of σ is not known; that is, use Student's t test here. Compare the results of the students with expectation.

16. The following data give the gains in weight of 15 chickens fed a special weight producing diet for a period of one month. Past experience for the standard diet over this period has yielded a mean gain of 9 ounces. Test the hypothesis, using a one-sided test, that the new diet is no improvement. The gains were: 12.5, 9.6, 10.8, 8.1, 11.4, 10.7, 9.0, 8.4, 7.5, 11.0, 8.7, 11.0, 10.3, 8.5, 9.0.

Section 3

17. Given two random samples of size 100 each from two normal populations with sample values $\bar{x}_1 = 20$, $\bar{x}_2 = 22$, $s_1 = 7$, $s_2 = 6$, test the hypothesis that $\mu_1 = \mu_2$.

18. Two sets of 50 elementary school children were taught to read by two different methods. After instruction was over, a reading test gave the following results: $\bar{x}_1 = 73.4$, $\bar{x}_2 = 70.2$, $s_1 = 9$, $s_2 = 10$. Test the hypothesis that $\mu_1 = \mu_2$.

19. In an industrial experiment a job was performed by 40 workmen according to method I and by 50 workmen according to method II. The results of the experiment yielded the following data on the length of time required to complete the job: $\bar{x}_1 = 54$ minutes, $\bar{x}_2 = 58$ minutes, $s_1 = 6$ minutes, $s_2 = 9$ minutes. Test the hypothesis that $\mu_1 = \mu_2$.

20. A flock of 100 turkeys was split into two groups of 50 each and fed two different rations. At the end of the experiment the following information was obtained: $\bar{x}_1 = 15.2$ lb., $s_1 = 2.2$ lb., $\bar{x}_2 = 14.8$, $s_2 = 2.0$. Test the hypothesis that $\mu_1 = \mu_2$.

21. Have each member of the class draw 10 pairs of one-digit random numbers (Table II in the appendix). The second number should be subtracted from the first number for each pair. Each student should bring to class the sum of the squares of these differences as well as the sum of the differences. By combining the class results, an estimate of σ_z^2, in which $z = x - y$, can be obtained and compared to the value expected from (4), namely $\sigma_{x-y}^2 = \sigma_x^2 + \sigma_y^2 = (2.87)^2 + (2.87)^2 = 16.5$. Here x and y denote, respectively, the first and second random digits of each pair. From earlier work it was found that $\mu_x = 4.5$ and $\sigma_x = 2.87$.

Section 3.1

▶**22.** Given two random samples of sizes 10 and 12 from two independent normal populations with $\bar{x}_1 = 20$, $\bar{x}_2 = 24$, $s_1 = 7$, $s_2 = 6$, test by means of the t distribution the hypothesis that $\mu_1 = \mu_2$, assuming that $\sigma_1 = \sigma_2$.

▶**23.** The following data give the gains of 20 rats, of which half received their protein from raw peanuts and the other half received their protein from roasted peanuts. Test by means of the t distribution to see whether roasting the peanuts had an effect on their protein value.

Raw	61	60	56	63	56	63	59	56	44	61
Roasted	55	54	47	59	51	61	57	54	62	58

▶**24.** Select two sets of 20 one-digit random numbers and test the hypothesis (which is true here) that the two population means are equal. Work first by large sample methods, using the fact that $\sigma = 2.87$; then work by means of Student's t distribution.

▶**25.** The following data give the gains in weight of 24 experimental animals that were split into two groups and fed two different rations. Test the hypothesis that $\mu_1 = \mu_2$, assuming that $\sigma_1 = \sigma_2$.

x_1	66	64	52	48	42	44	54	58	62	74	58	54
x_2	82	50	74	58	52	60	68	50	58	72	58	66

Section 4

26. If you rolled a die 240 times and obtained 52 sixes, would you decide that the die favored sixes?

27. Past experience has shown that 40 per cent of students fail a university entrance examination in English. If 54 out of 120 students from a certain city failed, would one be justified in concluding that the students from this city are inferior in English?

28. A biologist has mixed a spray designed to kill 50 per cent of a certain type of insect. If a spraying of 200 such insects killed 120 of them, would you conclude his mixture was satisfactory?

29. What is the difference between saying that a coin is honest for all practical purposes and saying that it is honest in a mathematical sense?

30. Explain why you might hesitate from practical considerations to reject the hypothesis that a coin is honest if you tossed the coin 1000 times and obtained 535 heads.

31. The following data were obtained for the daily percentage defective of parts for a production averaging 1000 parts a day. Construct a control chart and indicate whether control seems to exist here. The data are for the percentage (not proportion) of defectives and are to be read a row at a time. Estimate p by calculating the mean of these sample values.

2.2 2.3 2.1 1.7 3.8 2.5 2.0 1.6 1.4 2.6 1.5 2.8 2.9 2.6 2.5
2.6 3.2 4.6 3.3 3.0 3.1 4.3 1.8 2.6 2.1 2.2 1.8 2.4 2.4 1.6
1.7 1.6 2.8 3.2 1.8 2.6 3.6 4.2

Section 5

32. In a poll taken among college students 44 of 200 fraternity men favored a certain proposition, whereas 51 of 300 nonfraternity men favored it. Is there a real difference of opinion on this proposition?

33. In a poll of the television audience in a city 56 out of 200 men disliked a certain program, whereas 75 out of 300 women disliked it. Is there a real difference of opinion here?

34. In one section of a city 65 out of 450 taxpayers were delinquent with their tax payments, whereas in another section 42 out of 500 were delinquent. Test to see if the delinquency rate is the same for those two sections of the city.

35. A test of 100 youths and 200 adults showed that 50 of the youths and 64 of the adults were careless drivers. Use these data to test the claim that the youth percentage of careless drivers is larger than the adult percentage by 10 percentage points against the alternative that it exceeds this amount.

36. Two classes of the same size are taught by the same instructor. One class is a regular class, whereas the other one is in an adjoining class room and receives the instructor's lecture by closed circuit television. Discuss possible factors that might prevent a valid statistical comparison to be made of the two learning methods.

CHAPTER 9

Correlation and Regression

$2\frac{7}{60} = \frac{9}{20}$

1. LINEAR CORRELATION

The statistical methods presented thus far have all been concerned with a single variable x and its distribution. In particular, the preceding two chapters have been concerned with the estimation of, and the testing of hypotheses about, the parameters of binomial and normal variable distributions. Many of the problems in statistical work, however, involve several variables. This chapter is devoted to explaining two of the techniques for dealing with data associated with two or more variables. The emphasis is on two variables, but the methods can be extended to deal with more than two.

In some problems the several variables are studied simultaneously to see how they are interrelated; in others there is one particular variable of interest, and the remaining variables are studied for their possible aid in throwing light on this particular variable. These two classes of problems are usually associated with the names of *correlation* and *regression,* respectively. Correlation methods are discussed first.

A correlation problem arises when an individual asks himself whether there is any relationship between a pair of variables that interests him. For example, is there any relationship between smoking and heart ailments, between music appreciation and scientific aptitude, between radio reception and sun-spot activity, between beauty and brains?

For the purpose of illustrating how one proceeds to study the relationship between two variables, consider the data of Table 1, which consist of high school and first year college grade point averages. The high school average is denoted by x and the college average by y.

TABLE 1

x	y	x	y	x	y
3.0	2.4	2.9	1.9	3.1	2.8
2.4	2.6	2.7	2.2	3.3	3.2
3.7	3.0	3.7	3.1	2.7	1.8
3.6	3.9	2.7	2.6	3.5	2.7
3.8	3.6	3.3	2.8	2.9	2.1
2.9	3.0	2.8	2.7	2.7	1.7
3.5	3.1	3.1	2.4	2.9	1.7
3.0	2.8	2.8	3.0	3.2	2.3
2.3	2.2	3.0	3.3	3.4	2.6
3.0	2.9	2.2	1.8	2.5	2.7

The investigation of the relationship of two variables such as these usually begins with an attempt to discover the approximate form of the relationship by graphing the data as points in the x, y plane. Such a graph is called a *scatter diagram*. By means of it, one can quickly discern whether there is any pronounced relationship and, if so, whether the relationship may be treated as approximately linear. The scatter diagram for the 30 points obtained from the data of Table 1 is shown in Fig. 1.

An inspection of this scatter diagram shows that there is a tendency for small values of x to be associated with small values of y and for large values of x to be associated with large values of y. Furthermore, roughly speaking, the general trend of the scatter is that of a straight line. In determining the nature of a trend, one looks to see whether there is any pronounced tendency for the points to be scattered on both sides of some smooth curve with a few waves or whether they appear to be scattered on both sides of a straight line. It would appear here that

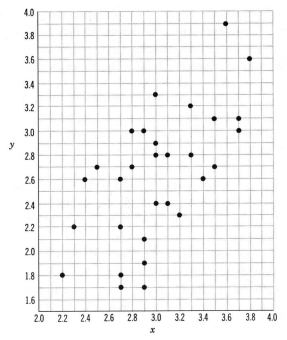

FIGURE 1.   Scatter diagram for grade point averages.

a straight line would do about as well as some mildly undulating curve. For variables such as these, it would be desirable to be able to measure in some sense the degree to which the variables are linearly related. For the purpose of devising such a measure, consider the properties that would be desirable.

A measure of relationship should certainly be independent of the choice of origin for the variables. The fact that the scatter diagram of Fig. 1 was plotted with the axes conveniently chosen to pass through the point (2.0, 1.5) implies that the relationship was admitted to be independent of the choice of origin. This property can be realized by using the deviations of the variables from their mean rather than the variables themselves. This was done in defining the standard deviation in Chapter 2. Thus one uses the variables $x_i - \bar{x}$ and $y_i - \bar{y}$ in place of the variables x_i and y_i in constructing the desired measure of relationship. The notation x_i, y_i denotes the ith pair of numbers in Table 1.

A measure of relationship should also be independent of the scale of measurement used for x and y. Thus, if the x and y scores of Table 1 were doubled the relationship between the variables should be unaffected thereby. Similarly, if one were interested in studying the relationship between stature of husbands and wives, one would not want the measure of the relationship to depend upon whether stature was measured in centimeters or inches. This property can be realized by dividing x and y by quantities that possess the same units as x and y. For reasons that will be appreciated presently, the quantities that will be chosen here are s_x and s_y, the two sample standard deviations. Both properties are therefore realized if the measure of relationship is constructed by using the variables x_i and y_i in the forms $u_i = (x_i - \bar{x})/s_x$ and $v_i = (y_i - \bar{y})/s_y$. This merely means that the x's and y's should be measured in sample standard units. This corresponds to doing for samples what was done in (4), Chapter 5, to measure a variable in theoretical standard units.

The scatter diagram of the points (u_i, v_i) for the data of Table 1 is shown in Fig. 2. It will be observed that most of the points are located in the first and third quadrants and that the points in those quadrants tend to have larger coordinates, in magnitude, than those in the second and fourth quadrants. A simple measure of this property of the scatter is the sum $\sum_{i=1}^{n} u_i v_i$. The terms of the sum contributed by points in the first and third quadrants will be positive, but those corresponding to points in the second and fourth quadrants will be negative. A large positive value of this sum would therefore seem to indicate a

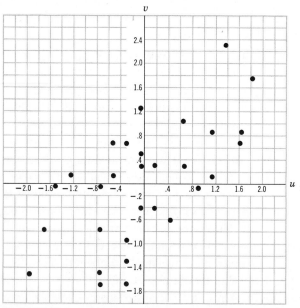

FIGURE 2. Scatter diagram for standard-ized grade point averages.

strong linear trend in the scatter diagram. This is not strictly true, however, for if the number of points were doubled without changing the nature of the scatter the value of the sum would be approximately doubled. It is therefore necessary to divide the sum by n, the number of points, before using it as a measure of relationship. There are theoretical reasons for preferring to divide by $n - 1$ rather than n here, just as in the case of defining the sample standard deviation. The resulting sum $\Sigma u_i v_i / (n - 1)$ is the desired measure of relationship. It is called the *correlation coefficient* and is denoted by the letter r; hence, in terms of the original measurements, r is defined by the following formula:

(1)   ***Correlation Coefficient.*** $$r = \frac{\sum_{i=1}^{n} (x_i - \bar{x})(y_i - \bar{y})}{(n - 1)s_x s_y}.$$

$$\boxed{\sum_{i=1}^{n} x_{i} y_{i}} - n \bar{x} \bar{y}$$

Calculations with the data of Table 1 will show that $r = .63$ for those data. In order to interpret this value of r and to discover what values of r are likely to be obtained for various types of relationships between x and y, a number of different scatter diagrams have been plotted and the corresponding values of r computed in Fig. 3. The first four diagrams correspond to increasing degrees, or strength, of linear relationship. The fifth diagram illustrates a scatter in which x and y are closely related but in which the relationship is not linear. This illustration points out the fact that r is a useful measure of the strength of the relationship between two variables only when the variables are linearly related.

If these diagrams could be viewed from the reverse side of the page, the scatters would tend to go downhill instead of uphill. Calculations of r for the resulting scatters would show that the new values of r would be the negatives of the old values. Thus the strength of the relationship is given by the magnitude of r, whereas the sign of r merely tells one whether the values of y tend to increase or decrease with x, with a positive value indicating that y tends to increase with x.

The diagrams of Fig. 3, together with the associated values of r, make plausible two properties of r, namely that the value of r must satisfy the inequalities

$$-1 \leq r \leq 1$$

and that the value of r will be equal to plus 1 or minus 1 if, and only if, all the points of the scatter lie on a straight line. These properties of r can be demonstrated to be correct by mathematical methods.

Multiplying out the parentheses in (1) and summing term by term will show that the numerator in (1) can be written as $\Sigma xy - n\bar{x}\bar{y}$, which is often convenient to use in calculating the value of r.

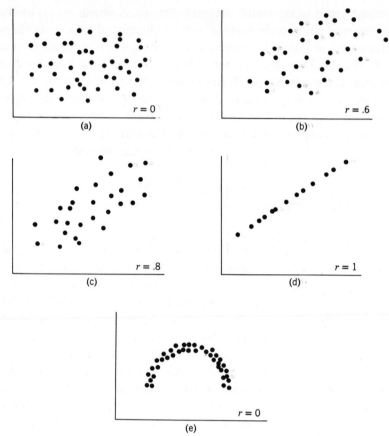

FIGURE 3.   Scatter diagrams and their associated values of *r*.

2.   INTERPRETATION OF *r*

The interpretation of a correlation coefficient as a measure of the strength of the linear relationship between two variables is a purely mathematical interpretation and is completely devoid of any cause or effect implications. The fact that two variables tend to increase or decrease together does not imply that one has any direct or indirect effect on the other. Both may be influenced by other variables in such a manner as to give rise to a strong mathematical relationship. For example, over a period of years the correlation coefficient between teachers' salaries and the consumption of liquor turned out to be .98. During this period of time there was a steady rise in wages and salaries of all types and a general upward trend of good times. Under such conditions, teachers' salaries would also increase. Moreover, the general upward trend in wages and buying power, together with the increase in population, would be reflected in increased total purchases of liquor. Thus the high correlation merely reflects the common effect of the upward trend on the two variables. Correlation coefficients must be handled with care if they are to give sensible information concerning relationships between pairs of variables. Success with them requires familiarity with the field of application as well as with their mathematical properties.

Correlation coefficients have proved very useful, for example, in psychological testing and in other fields where it is important to determine the interrelationships between several variables that are being studied simultaneously. Thus, the correlations between college grade point averages, high school grade point averages, aptitude test scores, vocabulary test scores, and similar variables have enabled some private colleges to evaluate the relative importance of such factors in determining academic success in college. Although the correlation between high school and college grade point averages for the students of Table 1 does not seem very high, it is typical of the value obtained in such studies. The magnitude of r depends much on the quality and the grading standards of the high schools from which the students came.

▶ 3.   RELIABILITY OF r

The value of r obtained for Table 1 may be thought of as the first sample value of a sequence of sample values, $r_1, r_2, r_3, \ldots$, that would be obtained if repeated sets of similar data were obtained. Such sets of data are thought of as having been obtained from drawing random samples of size $n = 30$ from a population of students. If the values of r were classified into a frequency table and the resulting histogram sketched, a good approximation would be obtained to the limiting, or theoretical, frequency distribution of r. As in the case of other variables, such as $\bar{x}$ and $\bar{x}_1 - \bar{x}_2$, the limiting distribution can be derived by mathematical methods if the proper assumptions concerning the variables x and y are made. If, for example, it is assumed that x and y are independent normal variables, in which case they are necessarily uncorrelated, the derivation shows that the desired sampling distribution of r depends only on n. Figure 4 shows the nature of this distribution when $n = 10$ and $n = 20$. By means of these distributions one

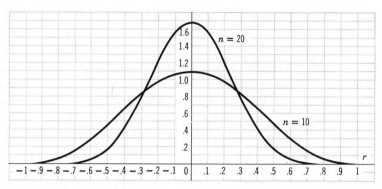

FIGURE 4. Distribution of r for $\rho = 0$ when $n = 10$ and $n = 20$.

can determine whether a sample value of r is large enough, numerically, to refute the possible claim that x and y are actually uncorrelated variables, that is, whether the theoretical correlation coefficient ρ, of which r is a sample estimate has the value 0. Table VI in the appendix gives critical values of r for these distributions, that is, values such that the probability is less than α that a sample value of r will exceed the critical value. The values of α selected are .05, .025, and .005. If there is no reason to believe that the correlation between x and y is positive, or negative, provided it is not zero, then a two-sided test should be used. Thus, the critical value of r corresponding to $\alpha = .025$ yields a two sided test with an .05 significance level. From Fig. 4 it will be observed, for example, that the .025 critical value for $n = 20$ should be somewhere between .4 and .5, which is in agreement with the Table VI value of .444. For a sample of size 20, therefore, the hypothesis of no correlation between x and y would be accepted unless the magnitude of r exceeded .444.

Since $n = 30$ and $r = .63$ for the data of Table 1, it follows from the Table VI value of .306 that a hypothesis of no correlation between high school and college grade point averages would be rejected here. The .05 column should be used in this problem because the alternative to zero correlation would certainly be positive correlation, and therefore only the right tail critical region should be selected.

From Table VI it should be apparent that a large sample is required before the claim that no correlation exists between two variables can be refuted, unless, of course, the value of r is very large. The moral here seems to be that the correlation coefficient is an interesting and often useful tool for studying the interrelationships between variables but is of questionable reliability and interpretation as a quantitative tool for analyzing those variables. In the next section a much more useful technique for studying two or more variables simultaneously will be introduced.

4.	LINEAR REGRESSION

Usually one studies the relationship between two or more variables in the hope that any relationship that is found can be used to assist in making estimates or predictions of a particular one of the variables. Thus, in studying the correlation between high school and college grade point averages, the intention obviously is to use that relationship to try to predict a college student's academic success from a knowledge of his high school grade point average. The correlation coefficient is merely concerned with determining how strongly two such variables are linearly related and it is not capable of solving prediction problems. Similarly, if correlation coefficients were calculated between college grade point averages and such variables as aptitude test scores and vocabulary test scores, those correlations would serve principally to indicate which variables are worth including in a prediction function for predicting academic success. Methods that have been designed to handle prediction problems are known as regression methods. This chapter discusses some of those methods.

For the purpose of explaining regression methods, consider the particular problem of predicting the yield of hay as a function of the amount of irrigation water applied. The data in Table 2 represent the amount of water applied in inches and the yield of alfalfa in tons per acre on an experimental farm. The graph of these data is given in Fig. 5. From this graph, it appears that x and y are approximately linearly related for this range of x values. A straight line will therefore be fitted to this set of points for the purpose of trying to predict the value of y from the value of x. Such a line has been fitted in Fig. 5. Now, for any given value of x, say $x = 30$, the predicted value of y is chosen as the distance up to the line directly above that value of x. Reading across on the y axis, it will be observed that the predicted value of y for $x = 30$ is approximately 7, as compared to the observed value of 7.21.

TABLE 2

Water (x)	12	18	24	30	36	42	48
Yield (y)	5.27	5.68	6.25	7.21	8.02	8.71	8.42

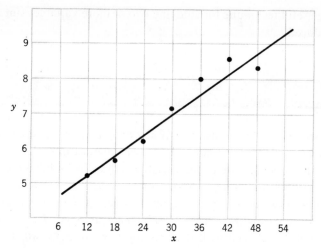

FIGURE 5. Hay yield as a function of amount of irrigation.

Suppose, now, that it is assumed that the relationship between mean yield and water is strictly linear over this range of x values. This means that if this experiment had been repeated a large number of times under the same growth conditions and if the y values corresponding to each of the seven x values had been averaged separately, then those averages would yield a set of points lying almost precisely on a straight line. The larger the number of such repetitions, the greater the expected precision. This assumption essentially says that there is a theoretical straight line that expresses the linear relationship between the theoretical mean value of y and the corresponding value of x.

If one accepts the linearity assumption, then one would expect the sample straight line value of approximately 7 to be closer to the theoretical line value for $x = 30$ than the observed value of 7.21 because one would expect the sample straight line, which is based on all seven experimental points, to be more stable than a single observed point. In view of this reasoning, one would predict the theoretical line value corresponding to $x = 30$ to be the corresponding y value on the sample regression line. Similar predictions would be made for the six other x values. Furthermore, if one were interested in an intermediate value of x, he would use the sample straight-line value as the predicted value of y for this value of x also. Since it is being assumed that the relationship is linear only for this range of x values, it is not legitimate to use the sample straight line to predict y values beyond this range of x values.

5.   LEAST SQUARES

In view of the preceding discussion, the problem of linear prediction reduces to the problem of fitting a straight line to a set of points. Now the equation of a straight line can be written in the form

$$y = a + bx,$$

in which a and b are parameters determining the line. Thus the equations

$$y = 2 + 3x$$

and

$$y = 4 - 2x$$

determine the two straight lines graphed in Fig. 6. The parameter a determines where the line cuts the y axis. Thus the two lines in Fig. 6 cut the y axis at 2 and 4, respectively. The parameter b determines the slope of the line. The slope 3 for the first line means that the line rises 3 units vertically for every positive horizontal unit change. A negative value such as -2 means that the line drops

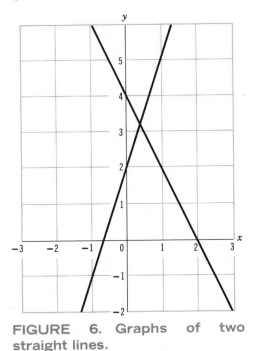

FIGURE 6. Graphs of two straight lines.

2 units for every positive horizontal change of 1 unit. Since only two points are needed to graph a line, these lines are readily graphed by assigning two different values to x and calculating the corresponding values of y to give the coordinates of two points on the line. To graph the two lines of Fig. 6, the following pairs of points were computed:

$y = 2 + 3x$		
x	0	1
y	2	5

$y = 4 - 2x$		
x	0	2
y	4	0

Since the problem is to determine the values of the parameters a and b so that the line will fit a set of points well, the problem is essentially one of estimating the parameters a and b in some efficient manner. Although there are numerous methods for performing the estimation of such parameters, the best known for regression problems is the *method of least squares*.

Since the desired line is to be used for predicting purposes, it is reasonable to require that the line be such that it will make the errors of prediction small. By an error of prediction is meant the difference between an observed value of y and the corresponding straight line value of y. For example, the error of prediction in Fig. 5 for $x = 30$ is approximately equal to $7.21 - 7 = .21$. If a different straight line had been used for prediction, this error of prediction ordinarily would have been different. Since the errors may be positive or negative and might add up to a small value for a poorly fitting line, it will not do to require merely that the sum of the errors be as small as possible. This difficulty can be avoided by requiring that the sum of the absolute values, that is, the magnitudes of the errors be as small as possible. However, sums of absolute values are not convenient to work with mathematically; consequently the difficulty is avoided by requiring that the sum of the squares of the errors be as small as possible. The values of the parameters that minimize the sum of the squares of the errors determine what is known as the best fitting straight line in the sense of least squares. It is clear from inspecting Figs. 5 and 6 that by varying a and b properly one should be able to find the equation of a line that fits the points of Fig. 5 well. The problem, however, is one of finding a best fitting line in some systematic rational way, and this is where the principle of least squares enters.

The problem of determining the least squares values of a and b requires more mathematical background than that expected for this book; therefore, it is not considered here. The results of applying the proper mathematical methods, however, are written down. It turns out that they are somewhat simpler if the variable x is measured from the sample mean. Thus the equation of the line is written in the form

(2) $$y' = a + b(x - \bar{x}).$$

The prime is placed on y only to distinguish the straight line value of y from the observed value of y for any x. When the equation of a straight line is written in form (2), b is still the slope of the line, but a is now the value of y when $x = \bar{x}$ rather than the value of y when $x = 0$. The least squares estimates of a and b for the equation of the line expressed in form (2) turn out to be

$$a = \bar{y} \quad \text{and} \quad b = \frac{\Sigma(x_i - \bar{x})y_i}{\Sigma(x_i - \bar{x})^2}.$$

In these formulas it is assumed that there are n pairs of values of x and y, such as those in Table 2, and that x_i and y_i denote the ith pair of values. Although the range of summation is not indicated on the Σ symbol, it is understood to be $i = 1$ to $i = n$. Thus, for Table 2, the summation would be $i = 1$ to $i = 7$. These values when substituted in (2) yield the desired least squares line. This line is customarily called the *regression line of y on x*; hence the preceding results may be summarized as follows:

(3)   ***Regression Line.*** $y' = \bar{y} + b(x - \bar{x}), \text{ where } b = \dfrac{\Sigma(x_i - \bar{x})y_i}{\Sigma(x_i - \bar{x})^2}.$

A pioneer in the field of applied statistics gave the least squares line this name in connection with some studies he was making on estimating the extent to which the stature of sons of tall parents reverts, or regresses, toward the mean stature of the population. That is, he found that sons of tall fathers tended to be less tall than the father, and sons of short fathers tended to be taller than the father.

For computational purposes, it is convenient to change the form of b slightly. If one multiplies out the factors in the numerator and simplifies the denominator in the same manner that s^2 was simplified in Chapter 2, it is easily shown that b reduces to

(4) $$b = \frac{\Sigma x_i y_i - n\bar{x}\bar{y}}{\Sigma x_i^2 - n\bar{x}^2}.$$

Table 3 illustrates the computational procedure for the data of Table 2. Here (3) was used to calculate b instead of the suggested computing formula (4) because $\bar{x}$ is an integer here, namely 30. Formula (3) then gives the values

$$a = \frac{49.56}{7} = 7.08 \quad \text{and} \quad b = \frac{103.68}{1008} = .103.$$

		TABLE 3		
x	y	$x - \bar{x}$	$(x - \bar{x})y$	$(x - \bar{x})^2$
12	5.27	−18	−94.86	324
18	5.68	−12	−68.16	144
24	6.25	−6	−37.50	36
30	7.21	0	0	0
36	8.02	6	48.12	36
42	8.71	12	104.52	144
48	8.42	18	151.56	324
210	49.56		103.68	1008

Consequently, the equation of the regression line is

$$y' = 7.08 + .103(x - 30).$$

Multiplying out the parenthesis and collecting terms will yield

$$(5) \qquad\qquad y' = 3.99 + .103x.$$

This is the equation of the regression line that was graphed in Fig. 5.

This problem illustrates very well the basic difference between correlation and regression methods for two variables. In the correlation problem corresponding to the scatter diagram of Fig. 1 the data consisted of a random sample of size 30 from a population of students. This means that both x and y are statistical variables whose values are determined only after the sample is obtained. In the preceding regression problem, however, the x values were chosen in advance, so that only the y values are determined by the sample. Now the least squares technique of fitting a line to a set of points can be applied whether the x values are fixed in advance or were obtained from random samples; hence regression methods could have been applied to data of the type considered in the study of correlation. On the other hand, the interpretation of r as a measure of the strength of the linear relationship between two variables obviously does not apply if the values of x are selected as desired, because the value of r will usually depend heavily upon the choice of x values.

In addition to being more flexible, regression methods also possess the advantage of being the natural methods to use in many experimental situations. The experimenter often wishes to change x by uniform amounts over the range of interest for that variable rather than to take a random sample of x values. Thus, if he wished to study the effect of an amino acid on growth, he would wish to increase the amount of amino acid by a fixed amount, or factor, each time he ran the experiment.

Although a correlation coefficient is useful for describing how strongly two variables are linearly related, it is not very useful otherwise. If the correlation coefficient between aptitude in mathematics and aptitude in music is .4, whereas that between aptitude in mathematics and aptitude in art is .2, one can conclude that the first association is stronger than the second and that both relationships are rather weak, but not a great deal more can be said. Correlation coefficients do not lend themselves readily to quantitative statements unless they are associated with regression. Thus correlation is usually only a first phase in the study of the relationship of two variables, whereas regression is the basic technique in such a study.

6.	THE REGRESSION FALLACY

The name of regression, which is commonly given to the least squares line, is associated with an error frequently made in the interpretation of observations taken at two different periods of time. This error, which is called the regression fallacy, is best explained by means of illustrations.

Suppose a teacher studies the scores of his students in a given course on their first two hour tests. If he selects, say, the top five papers from the first test and calculates the mean score for those five students on both tests, he will very likely discover that their second test mean is lower than their first test mean. Similarly, if he calculates the mean score on both tests for the five students having the lowest scores on the first test, he will undoubtedly find that their mean has risen. Thus he might conclude that the good students are slipping, whereas the poor students are improving. The explanation lies partly in the reaction of students to test scores but also partly in the natural variability of students' test scores. Even if students did not vary from one test to another in their study habits and in their total relative knowledge of the subject, the inaccuracy of a test to measure this knowledge would cause considerable variation in a student's performance from test to test. As a result, some of the top five students on the first test may be there because of fortuitous circumstances; however, the second test is likely to bring them down to their natural level, thereby dragging the mean of this group of five students down with them. The same type of reasoning applies to the lowest five students on the first test to account for their improved mean.

There is undoubtedly a psychological factor operating here also to accentuate the "regression" of high and low scores toward the mean of the entire group. Students who did poorly on the first test would be expected to study considerably harder than before and thus raise their scores on the second test. Since early success often leads to overconfidence, the students making the highest scores on the first test might be expected to ease up slightly and thus lower their mean score on the second test. Nevertheless, the natural variation of test scores alone will suffice to produce a fair amount of regression toward the group mean.

The regression fallacy often occurs in the interpretation of business data. For example, in comparing the profits made by a group of similar business firms for two consecutive years, there might be a temptation to claim that the firms with high profits are becoming less efficient, whereas those with low profits are becoming more efficient because the mean profit for each of those two extreme groups would tend to shift toward the mean of the entire group. The firms with high profits the first year were high in the list either because they are normally highly efficient or because they were fortunate that year but are normally of lower efficiency. The latter group would be expected to show lower profits the second year and thus decrease the mean for the first year's high profit group. The same type of reasoning would explain the apparent increased efficiency of the first year's low profit group.

The original study of the relationship between the stature of fathers and sons, which gave rise to the name regression, is another illustration of this type of possible misinterpretation. It was found that the tallest group of the total group of men being studied had sons whose mean height was lower than that of the fathers. It was also found that the shortest group had sons whose mean height was higher than that of the fathers. As in the other illustrations, the explanation lies in the natural variation of subgroups of a population. Since many tall men come from families whose parents are of average size, such tall men are likely to have sons who are shorter than they are; consequently, when a group of tall men is selected, the sons of such men would not be expected to be quite so tall as the fathers. There are, of course, factors such as the tendency of tall men to marry taller-than-average women and the steady increase in stature from generation to generation to dampen the above regression tendency somewhat.

7.   STANDARD ERROR OF ESTIMATE

After a regression line has been fitted to a set of points, it is usually possible to inspect its graph, such as the one in Fig. 5, and observe how accurately it predicts y values. An arithmetical procedure for doing this is to calculate the sizes of all the errors, $y_i - y_i'$, by reading them off the graph, if it is on graph paper, or by computing them by means of formula (3). A useful measure of the accuracy of prediction is obtained by calculating the standard deviation of the errors of prediction. If the error of prediction $y_i - y_i'$, corresponding to $x = x_i$, is denoted by e_i, then by definition the standard deviation of the errors is

$$\sqrt{\frac{\sum_{i=1}^{n} (e_i - \bar{e})^2}{n - 1}}.$$

But it can be shown that $\bar{e} = 0$. This means that the positive errors corresponding to points above the regression line cancel the negative errors corresponding to points below the regression line. In view of this property, the standard deviation of those errors will reduce to

$$\sqrt{\frac{\sum_{i=1}^{n} e_i^2}{n - 1}}.$$

For problems related to finding confidence limits for the parameters a and b, it turns out that it is better to divide the sum of squares of the errors by $n - 2$ rather than by $n - 1$, just as it was considered better to divide by $n - 1$ rather than by n in defining the ordinary sample variance. The same type of arguments apply here as applied there. It can be shown that the result is an unbiased estimate of σ^2, where σ^2 is the variance of the errors of prediction when the y_i' values are those of a theoretical regression line rather than of the sample regression line. The square root of the resulting expression is denoted by s_e and is called the *standard error of estimate*. Thus, in terms of the original variables,

(6)   ***Standard Error of Estimate.*** $s_e = \sqrt{\dfrac{\sum\limits_{i=1}^{n} (y_i - y_i')^2}{n - 2}}.$

Table 4 gives the calculation of s_e for the data of Table 2 by means of formulas (5) and (6).

If one assumes, as before, that there is a theoretical regression line, of which the least-squares line is an estimate, and in addition one assumes that the values of $y_i - y_i'$, where y_i' is now the theoretical line value, are independently and normally distributed with zero means and the same standard deviation σ, then s_e given by (6) is an estimate of σ. Furthermore, the normal distribution assump-

TABLE 4

y	y'	$y - y'$	$(y - y')^2$
5.27	5.23	.04	.0016
5.68	5.84	−.16	.0256
6.25	6.46	−.21	.0441
7.21	7.08	.13	.0169
8.02	7.70	.32	.1024
8.71	8.32	.39	.1521
8.42	8.93	−.51	.2601
			.6028

$$s_e = \sqrt{\frac{.6028}{5}} = .35$$

tion enables one to make approximate probability statements about the errors of prediction. For example, one can state that approximately 95 per cent of the errors of prediction will be less than $1.96s_e$ in magnitude. The approximation arises because 1.96σ has been replaced by its sample estimate $1.96s_e$ and because only the sample regression line is available; therefore, this is a large sample technique. Since most regression problems involve small samples, particularly the experimental kind like the one in Table 2, this technique is of somewhat limited use. A small sample method for dealing with regression problems is discussed in the next section.

Even though the sample given by Table 2 is too small to justify the large sample methods that have just been discussed, the results of the calculations that yielded equation (5) and the value .35 for (6) will be used to sketch the type of graph that one might legitimately construct if one had a larger sample. For this purpose it suffices to calculate the number $1.96s_e = 1.96(.35) = .7$ and then draw two lines parallel to the line (5) which is shown in Fig. 5, one being .7 unit above that line and the other being .7 unit below it. These two lines are shown in Fig. 7.

From Fig. 7 it will be seen that all seven points lie within the 95 per cent band, which of course was to be expected. In the long run of similar experiments one would expect approximately 95 per cent of such points to lie inside the band constructed in this fashion, provided the normal distribution assumption is satisfied.

This geometrical manner of looking at the prediction problem is a useful one for giving the experimenter a rough idea of what y values he is likely to obtain if he performs experiments at other x values.

The problem of predicting a y value for an x value beyond the range of observed

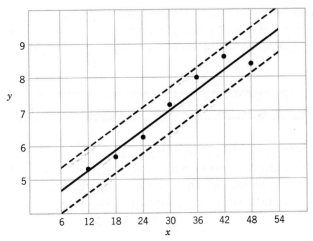

FIGURE 7. An approximate 95 per cent prediction band.

x values is a considerably more difficult problem than that of predicting for a value of x inside the interval of observations. Prediction beyond the range of observations is called extrapolation, whereas prediction inside the range is called interpolation. The difficulty with extrapolation is that the assumptions necessary to justify it are seldom realized in real-life situations. Thus, it is highly unlikely that the relationship between the amount of irrigation water applied and the yield of hay will continue to be linear beyond the range of x values given in Fig. 1. It certainly will not be if x is chosen sufficiently large. Furthermore, in most realistic regression problems it is unreasonable to expect the prediction errors to be normally distributed about the regression line with the standard deviation of those errors remaining constant as x goes beyond the interval of observations. Extrapolation is a legitimate technique only when the experimenter has valid reasons for believing that his model holds beyond the range of the available observations. Stock market prices (y) as a function of time (x), for example, are an excellent illustration of a regression problem for which no satisfactory extrapolation model has yet been constructed.

A kind word needs to be said for the usefulness of straight line regression models in realistic situations. If the scatter diagram for a set of data indicates that the relationship between x and y is not linear, it may still be possible to use the linear model if one can find a function of x and also a function of y such that the relationship between those two functional values is linear. There are many known relationships in science that are not linear but which can be made linear by taking the proper functions of x and y. This technique of considering the relationship between functions of x and y rather than between x and y extends the range of applicability of linear methods considerably.

As indicated in the preceding paragraph, in order to be able to make probability statements about errors of prediction, it is necessary to assume that there exists a theoretical regression line, of which the least squares line is a sample estimate, and then make some distribution assumptions. This postulated line is written in the form

$$y' = \alpha + \beta(x - \bar{x}).$$

Here α and β play the same role for this theoretical line as a and b do for the line determined by the sample points and given by (2). Thus the values of a and b given by (3) are the estimates of α and β obtained by least squares. There are no new problems of estimation or hypothesis testing in connection with the parameter α because, from (3), its estimate is $\bar{y}$ and therefore the small sample methods for treating a normal mean, explained in earlier chapters, may be applied here. The slope parameter β, however, does require new formulas. Fortunately, it can be shown that Student's t distribution can be applied to problems related to β. In particular, it can be shown that the quantity

(7)
$$t = \frac{b - \beta}{s_e} \sqrt{\Sigma (x_i - \bar{x})^2}$$

possesses a Student's t distribution with $\nu = n - 2$ degrees of freedom, provided the $y_i - y_i'$ are independently and normally distributed with zero means and equal standard deviations.

For the purpose of illustrating how to use formula (7) consider the problem of finding a 95 per cent confidence interval for β for the problem of section 4. Since $\nu = 5$ for that problem, the .05 critical value of t in Table V in the appendix is $t = 2.57$. Therefore, the probability is .95 that t will satisfy the inequalities

$$-2.57 < t < 2.57.$$

From (7), this is equivalent to

$$-2.57 < \frac{b - \beta}{s_e} \sqrt{\Sigma (x_i - \bar{x})^2} < 2.57.$$

When these inequalities are solved for β, it will be found that they are equivalent to

(8) $\qquad b - 2.57 \dfrac{s_e}{\sqrt{\Sigma\,(x_i - \bar{x})^2}} < \beta < b + 2.57 \dfrac{s_e}{\sqrt{\Sigma\,(x_i - \bar{x})^2}}\,.$

If 2.57 is replaced by the proper t value, these inequalities may serve as a formula for finding confidence limits for β for any size sample and any desired percentage limits. Earlier calculations yielded the values $b = .103$ and $s_e = .35$. The last column of Table 3 yields the value $\Sigma\,(x_i - \bar{x})^2 = 1008$; hence $\sqrt{\Sigma\,(x_i - \bar{x})^2} = 31.7$. If these values are substituted in (8), it will reduce to

$$.103 - 2.57\dfrac{.35}{31.7} < \beta < .103 + 2.57\dfrac{.35}{31.7}$$

or

$$.075 < \beta < .131.$$

As a result, the desired 95 per cent confidence interval for β is the interval from .075 to .131.

Tests of hypotheses about β can be made by means of formula (7) in the usual manner.

▶ 9. MULTIPLE LINEAR REGRESSION

Methods for dealing with problems of predicting one variable by means of several other variables, rather than by means of just one other variable, are similar to those for one variable. For example, if one were to predict the variable y in terms of the two variables x_1 and x_2, the problem would become one of finding the best fitting plane, in the sense of least squares, to a scatter diagram of points in three dimensions. The geometry of such a problem is illustrated in Fig. 8.

Since the equation of any plane can be written in the form

(9) $\qquad\qquad\qquad y' = a_0 + a_1 x_1 + a_2 x_2,$

the problem is one of estimating the three parameters a_0, a_1, and a_2 by the method of least squares. This is done by mathematical methods in the same manner as for simple linear regression. It turns out that the least squares values of a_0, a_1, and a_2 are obtained by solving the following set of three linear equations:

(10)
$$
\begin{aligned}
a_0 n + a_1 \Sigma x_1 + a_2 \Sigma x_2 &= \Sigma y \\
a_0 \Sigma x_1 + a_1 \Sigma x_1{}^2 + a_2 \Sigma x_1 x_2 &= \Sigma x_1 y \\
a_0 \Sigma x_2 + a_1 \Sigma x_1 x_2 + a_2 \Sigma x_2{}^2 &= \Sigma x_2 y.
\end{aligned}
$$

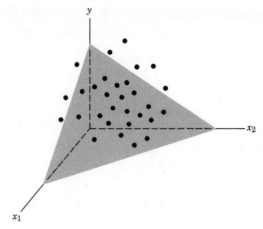

FIGURE 8. Regression plane in three dimensions.

This result generalizes for additional variables. Thus, if one had four variables, x_1, x_2, x_3, and x_4, by means of which to predict y, there would be five equations in the five unknowns a_0, a_1, a_2, a_3, and a_4 to solve.

One of the attractive features of the least squares approach in regression problems is the simple manner of writing down the equations that need to be solved to obtain estimates of the regression equation parameters. Thus, equations (10) can be written down by carrying out the following operations. Remove the prime on y in the multiple regression equation (9) so that it will represent an observed value of y rather than a regression value, then sum both sides of the resulting equation to obtain the first least-squares equation. Next, multiply both sides of this modified regression equation by x_1 and sum both sides to obtain the second least squares equation. Finally, multiply both sides by x_2 and sum to obtain the third equation.

If there had been five x variables instead of two in the regression equation, one would have continued this procedure by multiplying by x_3 and summing, then by x_4 and summing, and finally by x_5 and summing to obtain three more least squares equations. There would then be six least-squares equations to be solved for estimates of the six unknown regression parameters.

Since the reader is not expected to be familiar with methods for solving sets of equations, the details of multiple linear regression methods are omitted. The purpose of this section is to point out that methods are available for the more general problem and that these methods are very similar to those discussed in the preceding sections.

▶10. NONLINEAR REGRESSION

The discussion in section 7 pointed out the possibility of reducing many nonlinear relationships between two variables x and y to linear ones by choosing the proper functions of x and y. Although this is always possible theoretically, the proper functions may be exceedingly complicated or one may not be able to determine what the proper functions are. It is then often more satisfactory to work with nonlinear relationships between x and y. The simplest curve to use as a regression model if a straight line will not suffice is a parabola. Its equation can be written in the form

$$(11) \qquad\qquad y' = a_0 + a_1 x + a_2 x^2.$$

The problem of fitting a curve of this type to a set of points in the x, y plane by least squares is similar to that of fitting a multiple regression equation to a set of points in three dimensions. As a matter of fact, the results that were obtained in the preceding section can be applied directly to this problem. It is merely necessary to choose $x_1 = x$ and $x_2 = x^2$ in equation (9) to obtain equation (11). As a result the least squares equations for (11) are obtained by making these same substitutions in the least squares equations given in (10). Thus, the equations whose solutions give the desired estimates for a_0, a_1, and a_2 in fitting the parabola (11) to a set of points are the following:

$$(12) \qquad \begin{aligned} a_0 n + a_1 \Sigma x + a_2 \Sigma x^2 &= \Sigma y \\ a_0 \Sigma x + a_1 \Sigma x^2 + a_2 \Sigma x^3 &= \Sigma xy \\ a_0 \Sigma x^2 + a_1 \Sigma x^3 + a_2 \Sigma x^4 &= \Sigma x^2 y. \end{aligned}$$

For the purpose of observing to what extent a parabola fits a set of points better than a straight line, it suffices to calculate the errors of prediction for both fitted curves and compare them. The standard error of estimate given by (6) may also be calculated for both curves and compared; however, the standard error of estimate for a parabola uses the divisor $n - 3$ in place of $n - 2$ in formula (6). If there is very little difference between the two calculated standard errors of estimate, not much is gained by using a parabola rather than a straight line for the regression curve.

These methods can be extended to polynomial curves of higher degree. All that one needs to do is to replace x_3 by x^3, x_4 by x^4, etc., in the more general forms of (9) and (10) to obtain the desired degree polynomial regression curve equation and its corresponding least squares equations for estimating the coefficients. Formula (6) for the standard error of estimate is modified by dividing by $n - s$, where s denotes the number of unknown parameters in (9) that are being estimated.

11. ADDITIONAL ILLUSTRATIONS

1.   The following pairs of numbers are the scores made by ten students on a language test (x) and a science test (y). (a) 1 Plot the scatter diagram and observe whether the relationship may be treated as being linear. (b) 1 Calculate the value of r. (c) 3 Test the hypothesis that $\rho = 0$. (d) 5 Find the equation of the least squares line. (e) 7 Calculate the errors of prediction. (f) 7 Calculate the standard error of estimate. (g) 7 Determine what percentage of the errors of prediction exceed $2s_e$.

x	56	60	64	82	76	72	74	66	64	86
y	60	68	60	74	80	84	80	72	62	82

(a)

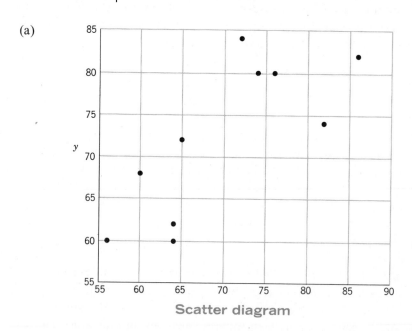

Scatter diagram

(b) Calculations give $\Sigma x = 700$, $\Sigma y = 722$, $\Sigma xy = 51{,}168$, $\Sigma x^2 = 49{,}840$, $\Sigma y^2 = 52{,}908$; hence

$$\Sigma xy - n\bar{x}\bar{y} = 628,$$

$$(n-1)s_x s_y = \sqrt{(n-1)s_x^2 \cdot (n-1)s_y^2} = \sqrt{[\Sigma x^2 - n\bar{x}^2][\Sigma y^2 - n\bar{y}^2]}$$

$$= \sqrt{[840][779.6]} = 809.2; \text{ hence } r = \frac{628}{809.2} = .78.$$

(c) Because a positive correlation is to be expected if there is no correlation, a one-sided test should be used. From Table VI the critical value for $n = 10$ is .549. Since $.78 > .549$, reject $H_0 : \rho = 0$.

(d) Using formula (4), $b = \dfrac{628}{840} = .75$; hence

$$y' = 72.2 + .75(x - 70) = 19.7 + .75x$$

(e)

x	56	60	64	82	76	72	74	66	64	86
y	60	68	60	74	80	84	80	72	62	82
y'	61.7	64.7	67.7	81.2	76.7	73.7	75.2	69.2	67.7	84.2
e	-1.7	3.3	-7.7	-7.2	3.3	10.3	4.8	2.8	-5.7	-2.2

(f) $\Sigma(y - y')^2 = 310.1$; hence $s_e = \sqrt{\dfrac{310.1}{8}} = 6.2.$

(g) $2s_e = 12.4$; hence none of the errors exceeds $2s_e$.

2. The following pairs of numbers were obtained by choosing three random digits from Table II and then adding the third digit to each of the first two to yield a pair labeled x and y. (a) 1 Plot the scatter diagram. Does the relationship appear to be linear? (b) 1 Calculate the value of r (c) 3 Test the hypothesis that $\rho = 0$.

x	8	8	4	10	11	1	13	12	6	14	7	8	6	13	8	4	7	8	5	14	5	10	8	13	17
y	7	8	11	14	13	3	8	8	3	8	3	11	5	11	5	10	11	3	14	13	2	10	17	10	15

x	14	9	7	10	13	9	5	13	12	5	9	6	12	14	0	10	10	17	9	13	12	9	1	12	12
y	7	11	10	1	12	2	5	13	14	14	15	11	10	15	5	13	3	8	11	9	10	14	5	14	12

(a)

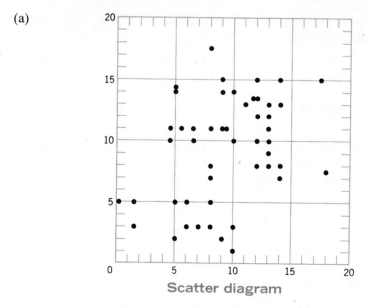

Scatter diagram

The relationship appears to be a fairly weak linear one.

(b) Calculations give $\Sigma x = 463$, $\Sigma y = 467$, $\Sigma x^2 = 5039$, $\Sigma y^2 = 5229$, $\Sigma xy = 4615$; hence using the formula $\Sigma xy - n\bar{x}\bar{y}$ for the numerator of (1)

$$r = \frac{4615 - (463)(9.34)}{\sqrt{\left[5039 - \dfrac{(463)^2}{50}\right]\left[5229 - \dfrac{(467)^2}{50}\right]}} = .36$$

(c) From Table VI the critical value is .279. Since $|r| > .279$, reject the hypothesis $H_0 : \rho = 0$.

3. Given the following data, (a) 4 plot the points, (b) 5 find the equation of the least squares line, (c) 5 graph the line on the graph containing the plotted points, (d) 5 predict the value of y for $x = 16.5$, (e) 7 calculate the errors of prediction, (f) 7 calculate the standard error of estimate, (g) 7 draw two lines parallel to the least squares line forming a band within which 68 per cent of the points might be expected to lie under a normal distribution assumption and comment, (h) 8 test the hypothesis that $\beta = .5$, (i) 8 find 90 per cent confidence limits for β, (j) 10 fit a parabola to the points, (k) 10 calculate the standard error of estimate for the fitted parabola and compare with the result in (f).

x	5	6	7	8	9	10	11	12	13	14	15	16	17
y	7.6	9.5	9.3	10.3	11.1	12.1	13.3	12.7	13.0	13.8	14.6	14.6	14.7

(a), (c), and (g)

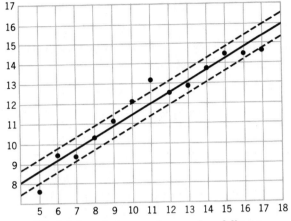

Plot of points and graph of lines

(b)

x	y	x^2	xy
5	7.6	25	38.0
6	9.5	36	57.0
7	9.3	49	65.1
8	10.3	64	82.4
9	11.1	81	99.9
10	12.1	100	121.0
11	13.3	121	146.3
12	12.7	144	152.4
13	13.0	169	169.0
14	13.8	196	193.2
15	14.6	225	219.0
16	14.6	256	233.6
17	14.7	289	249.9
$\overline{143}$	$\overline{156.6}$	$\overline{1755}$	$\overline{1826.8}$

$$\bar{x} = \frac{143}{13} = 11$$

$$\bar{y} = \frac{156.6}{13} = 12.05$$

$$b = \frac{1826.8 - 143(12.05)}{1755 - 143(11)} = .570$$

$$y' = 12.05 + .570(x - 11), \text{ or}$$
$$y' = 5.78 + .570x$$

Calculating formula (4) was used here even though it would have been easier to take advantage of the fact that $\bar{x} = 11$.

(d) $y' = 15.2$ for $x = 16.5$.

(e)

y	7.6	9.5	9.3	10.3	11.1	12.1	13.3	12.7	13.0	13.8	14.6	14.6	14.7
y'	8.6	9.2	9.8	10.3	10.9	11.5	12.0	12.6	13.2	13.8	14.3	14.9	15.5
e	-1.0	.3	$-.5$	.0	.2	.6	1.3	.1	$-.2$	.0	.3	$-.3$	$-.8$

Each value of y' was obtained from the preceding one by adding .57 to it. The first value was obtained by substituting $x = 5$ in the equation.

(f) From part (e), $\Sigma e^2 = 4.30$; hence $s_e = \sqrt{4.30/11} = .63$.

(g) There are four points lying outside this band, which is about what is to be expected.

(h) Using the fact that $\bar{x} = 11$, $\Sigma (x_i - \bar{x})^2 = 182$,

$$t = \frac{.57 - .50}{.63} \sqrt{182} = 1.50; \text{ hence accept } H_0:\beta = .5.$$

(i)
$$.57 - 1.796 \frac{.63}{\sqrt{182}} < \beta < .57 + 1.796 \frac{.63}{\sqrt{182}}, \text{ or}$$
$$.57 - .084 < \beta < .57 + .084, \text{ or}$$
$$.49 < \beta < .65.$$

(j)

x	y	x^3	x^4	$x^2 y$
5	7.6	125	625	1,900
6	9.5	216	1,296	3,420
7	9.3	343	2,401	4,557
8	10.3	512	4,096	6,592
9	11.1	729	6,561	8,991
10	12.1	1,000	10,000	12,100
11	13.3	1,331	14,641	16,093
12	12.7	1,728	20,736	18,288
13	13.0	2,197	28,561	21,970
14	13.8	2,744	38,416	27,048
15	14.6	3,375	50,625	32,850
16	14.6	4,096	65,536	37,376
17	14.7	4,913	83,521	42,483
		23,309	327,015	233,668

Equations (12) then become

$$a_0 13 + a_1 143 + a_2 1{,}755 = 156.6$$
$$a_0 143 + a_1 1{,}755 + a_2 23{,}309 = 1{,}826.8$$
$$a_0 1{,}755 + a_1 23{,}309 + a_2 327{,}015 = 23{,}366.8$$

or

$$a_0 + a_1 11 + a_2 135 = 12.046$$
$$a_0 + a_1 12.273 + a_2 163 = 12.775$$
$$a_0 + a_1 13.281 + a_2 186.33 = 13.314$$

or

$$a_1 1.273 + a_2 28 = .729, \ a_1 2.281 + a_2 51.33 = 1.268$$

or

$$a_1 + a_2 21.995 = .573, \ a_1 + a_2 22.505 = .556$$

or

$$a_2 = -.033, \ a_1 = 1.299, \ a_0 = 2.212$$

Hence

$$y' = 2.21 + 1.30x - .033x^2.$$

The following graph is that of this parabola.

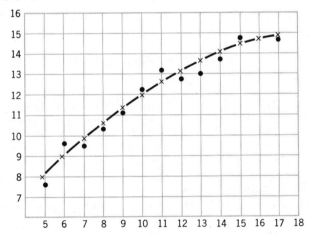

y	7.6	9.5	9.3	10.3	11.1	12.1	13.3	12.7	13.0	13.8	14.6	14.6	14.7
y'	7.9	8.8	9.7	10.5	11.2	11.9	12.5	13.1	13.5	13.9	14.3	14.6	14.8
e	$-.3$	.7	$-.4$	$-.2$	$-.1$	.2	.8	$-.4$	$-.5$	$-.1$	.3	.0	$-.1$

$s_e = \sqrt{\Sigma e^2 / 10} = .45$. Since $s_e = .63$ for the straight line, the parabola gives a better fit. The standard error has been decreased about $\frac{1}{3}$.

4. Given the following data, (a) 9 find the equation of its least squares plane, (b) 9 calculate the errors of prediction, (c) 9 calculate the standard error of estimate, (d) 5 omit the variable x_2 and then find the equation of the least squares line, (e) 7 calculate the standard error of estimate for the line obtained in (d) and compare with the value obtained in (c). Was much gained by introducing the variable x_2 in addition to x_1?

x_1	1	7	6	2	5	2	8	3	7	6
x_2	4	6	8	0	1	5	8	8	3	0
y	11	21	23	7	13	18	30	18	21	20

(a) Calculations give $\Sigma x_1 = 47, \Sigma x_2 = 43, \Sigma y = 182, \Sigma x_1^2 = 277, \Sigma x_2^2 = 279,$ $\Sigma x_1 x_2 = 218, \Sigma x_1 y = 972, \Sigma x_2 y = 904$; hence the following equations need to be solved

$$a_0 10 + a_1 47 + a_2 43 = 182$$
$$a_0 47 + a_1 277 + a_2 218 = 972$$
$$a_0 43 + a_1 218 + a_2 279 = 904$$

The solution of these equations is $a_0 = 5.48, a_1 = 1.81, a_2 = .98$; hence the equation of the regression plane is

$$y' = 5.48 + 1.81 x_1 + .98 x_2.$$

(b)

y	11	21	23	7	13	18	30	18	21	20
y'	11.2	24.0	24.2	9.1	15.5	14.0	27.8	18.8	21.1	16.3
e	-.2	-3.0	-1.2	-2.1	-2.5	4.0	2.2	-.8	-.1	3.7

(c) $\Sigma e_i^2 = 56.32$; hence $s_e = 2.84$.

(d) From (a), $\bar{y} = 18.2, b = 2.08$; hence $y' = 8.42 + 2.08 x$.

y	11	21	23	7	13	18	30	18	21	20
y'	10.5	23.0	20.9	12.6	18.8	12.6	25.1	14.7	23.0	20.9
e	.5	-2.0	2.1	-5.6	-5.8	5.4	4.9	3.3	-2.0	-.9

Here $\Sigma e^2 = 142.5, s_e = \sqrt{142.5/8} = 4.22$. Hence it follows from (c) where $s_e = 2.84$ that x_2 was very useful because it reduced the value of s_e by about $\frac{1}{3}$.

EXERCISES

Section 1

1. For the following data on the heights (x) and weights (y) of 12 college students, (a) plot the scatter diagram, (b) guess the value of r, (c) calculate the value of r.

x	65	73	70	68	66	69	75	70	64	72	65	71
y	124	184	161	164	140	154	210	164	126	172	133	150

2. Calculate the value of r for the following data on intelligence test scores and grade-point averages, after first plotting the scatter diagram and guessing the value of r.

I.T.	295	152	214	171	131	178	225	141	116	173	230
G.P.A.	3.4	1.6	1.2	1.0	2.0	1.6	2.0	1.4	1.0	3.6	3.6

	195	174	236	198	217	143	135	146	227
	1.0	2.8	2.8	1.8	2.0	1.2	2.4	2.2	2.4

3. Why is the value of the correlation coefficient unaffected if the variables x and y are interchanged?

Section 2

4. What would you guess the value of r to be for the following pairs of variables: (a) the number of man-hours of work and the number of units of a product produced in a given industry, (b) the size of a city and the crime rate, (c) the cost per unit of producing an article and the number of units produced?

5. Guess the value of r for the following pairs of variables: (a) mathematics and foreign language grades, (b) consumption of butter and the price of butter, (c) amount of rain in the spring and the mean temperature.

6. What interpretation would you give if told that the correlation between the number of automobile accidents per year and the age of the driver is $r = -.60$ if only drivers with at least one accident are considered?

7. Explain why it would not be surprising to find a fairly high correlation between the density of traffic on Wall Street and the height of the tide in Maine if observations were taken every hour from 6:00 A.M. to 10:00 P.M. and high tide occurred at 7:00 A.M. Plot a scatter diagram of relative values of these two variables for each hour of the day to assist you in the explanation.

8. For the data of problem 2, delete those items for which the intelligence test score is less than 150 and more than 225. Now calculate the value of r and compare with the value obtained for problem 2. What does this comparison seem to indicate?

Ch 8 glazer & appendix

9. Form 30 pairs of numbers by choosing four random digits from Table II and adding the last two to each of the first two to yield a pair labeled x and y. Work the problems of review exercise 2 in section 11. Do you believe that the value of ρ here is close to $\frac{1}{2}$? If not what would you guess its value to be?

Section 3

►**10.** Test the hypothesis that $\rho = 0$ if a sample of size 25 gave $r = .35$.

►**11.** The correlation between aptitude in mathematics and foreign languages, as based on tests purporting to measure such aptitudes, is found to be near .40. How large a sample of students should be taken to be quite certain that the value of r obtained would refute the claim that $\rho = 0$?

►**12.** Test the hypothesis $H_0: \rho = 0$ against $H_1: \rho > 0$ if a sample of size 40 gave $r = .30$.

►**13.** Test the hypothesis that $\rho = 0$ if a sample of size 30 gave $r = -.35$. Would you have arrived at the same conclusion if you had tested $H_0: \rho = 0$ against $H_1: \rho < 0$?

Section 5

14. Graph the line whose equation is $y = .3x - 1$ for values of x between 0 and 5.

15. Graph the line whose equation is $y = -.5x + 1.5$ for values of x between -2 and 4.

16. The regression line for estimating the yearly family expenditure on food by means of yearly income, in dollars, is given by $y = 200 + .20x$. (a) What is the average expenditure for families with incomes of 2000 dollars, 5000 dollars? (b) Why would you hesitate to use this formula for incomes of 0 dollars?

17. Find the equation of the regression line for the following data. Graph the line on the scatter diagram.

x	1	2	3	4	5
y	2	5	4	8	6

18. The following data are for tensile strength and hardness of die cast aluminum. Find the equation of the regression line for estimating tensile strength from hardness and graph it on the scatter diagram.

T.S. (y)	293	349	368	301	340	308	354	313	322	334	377	247
H (x)	53	70	84	55	78	64	71	53	82	67	70	56

19. The following data are the scores made by students on an entrance examination (x) and a final examination (y). (a) Find the equation of the regression line. (b) Graph the line on the scatter diagram.

x	129	179	347	328	286	256	477	430	327	245	286	326
y	370	361	405	302	496	323	374	332	435	165	375	466

20. Give examples of pairs of variables for which the regression line would be expected to have a negative slope.

Section 6

21. Explain the regression fallacy in the statement that track stars seem to go down hill after establishing a record.

22. Give an illustration of data for which the regression fallacy might easily be made.

23. Explain the regression fallacy in the statement that sophomore football stars seldom do as well in their junior year as they did in their sophomore year.

Section 7

24. Calculate the standard error of estimate in problem 17.

25. Calculate the standard error of estimate in problem 18 and find what percentage of the errors exceed it. What percentage of the errors exceed twice the standard error of estimate?

26. Explain why the standard error of estimate calculated in problem 24 and also in problem 25 would not be reliable for predicting the sizes of future errors of estimate.

Section 8

▶**27.** Given $n = 8$, $b = 2$, $s_e = .2$, and $\Sigma(x_i - \bar{x})^2 = 50$, find 95 per cent confidence limits for β.

▶**28.** Find 95 per cent confidence limits for β in problem 18.

▶**29.** Test the hypothesis that $\beta = 1.0$ in problem 17.

▶**30.** Test the hypothesis that $\beta = 2.5$ in problem 18.

▶**31.** By inspecting formula (8), determine how you would choose ten values of x between $x = 0$ and $x = 10$ at which to take observations on y so as to make the confidence interval for β as short as possible.

Section 9

32. Given the following data, use formulas (10) to find the equation of the linear regression function of y on x_1 and x_2.

x_1	0	1	2	3	4	5	6	7	8	9
x_2	2	10	6	12	4	11	15	14	9	13
y	0	4	9	18	10	16	20	20	18	25

33. For the data of problem 32, (a) find the equation of the linear regression function of y on x_1, (b) calculate the standard error of estimate and compare it with the standard error of estimate based on the function found in problem 32 to see whether the variable x_2 was beneficial.

Section 10

34. The following data give the number of man-hours (y) required to complete a job as a function of the number of units already completed. (a) Graph the data. (b) Fit a parabola to the set of points in (a) by means of formulas (12).

x	0	1	2	3	4	5
y	40	34	30	24	22	20

Explain why a parabola is an unrealistic type of curve to use as a model on a problem of this type.

Section 11

35. Given the following data, work parts (a) through (g) for the first of the review exercises of section 11.

x	1	9	6	4	5	7	7	1	2	8	6	3
y	15	75	55	42	33	45	55	17	32	80	48	45

▶**36.** Given the following data, work the problems in the fourth of the review exercises of section 11.

x_1	2	1	5	8	7	2	1	3	0	9	4	6
x_2	6	5	5	7	3	1	8	2	6	7	7	9
y	13	9	15	16	21	9	15	11	12	30	19	22

▶Special Topics

CHAPTER 10

Statistical Decision

1. INTRODUCTION

In the preceding chapters the emphasis has been largely on statistical inference, with very little attention to interpretation and decision making. That is because the action to be taken after a statistical inference has been made usually depends on outside considerations, economic or other, which are not available to the statistician. The purchaser of television tubes who has obtained sample estimates on the length of life of two competing brands of tubes, for example, must consider differences in price as well as quality in deciding which brand to buy.

There are certain types of problems, however, where the statistician can indicate a procedure for making an economically sound decision, provided the economic basis on which it is based is accepted. The simplest such problem is one in which there are only two possible states or events to consider and two possible actions available. For example, a truck gardener may be concerned about the possibility of a heavy freeze ruining his lettuce crop and must decide whether to take out insurance at the beginning of the season or to gamble on a freeze not occurring.

The two possible states or events here are that of a freeze occurring and no freeze occurring, and the two possible actions are taking out insurance and not doing so. A more general version of this type of problem occurs when there are more than two possible events and more than two possible actions. For example, the preceding problem could be generalized slightly by considering three possible actions: buying no insurance, buying enough insurance to cover half the possible losses, and buying insurance to cover all the possible losses. Problems involving more than two states and actions are solved in the same manner as those involving only two such; therefore, the discussion will be limited to this simple situation.

2. THE PAYOFF TABLE

For the purpose of explaining the reasoning behind a proposed solution for problems of the preceding type, consider the following particular problem. Suppose the truck gardener of the foregoing illustration will net $30,000 from his lettuce crop if no freeze occurs but that he will net only $10,000 if it does. This is, of course, a simplified version of the real situation because his losses will normally depend on the severity of the freeze. Suppose further that a $15,000 insurance policy will cost him $5000. Using net profit as a basis, all this information can be displayed in a two-by-two table, called a payoff table, as shown in Table 1.

TABLE 1

		Events	
		Freeze	No freeze
Action	Insurance	20,000	25,000
	No insurance	10,000	30,000

The basis for preferring one of these actions, or decisions, over the other will be that of expected profit. Thus, the decision to take out insurance will be made if the expected profit in doing so is larger than the expected profit when no insurance is purchased. Let p be the probability that a freeze will occur during the growing season. Then using only the first row entries of Table 1, it follows from the definition of expected value given by (6), Chapter 4, that the expected profit under insurance, which will be denoted by $E(I)$, is given by

$$E(I) = 20,000\,p + 25,000\,(1 - p).$$

If no insurance is purchased this expectation, denoted by $E(N)$, becomes

$$E(N) = 10,000\,p + 30,000\,(1 - p).$$

Assume for the present that the value of p is unknown. Then the two possible decisions will be equally good provided p satisfies the equation obtained by setting $E(I) = E(N)$. The value of p that accomplishes this, that is, the value of p that produces equality between the two expected values of a problem is called the *break even* point. For the problem under discussion the break even point is the value of p satisfying the equation

$$20,000\,p + 25,000\,(1 - p) = 10,000\,p + 30,000\,(1 - p).$$

Dividing through by 1000 will reduce this equation to

$$20\,p + 25\,(1 - p) = 10\,p + 30\,(1 - p)$$

which in turn is equivalent to

$$10\,p = 5\,(1 - p).$$

The solution of this equation is $p = \frac{1}{3}$. Consequently, if $p > \frac{1}{3}$ insurance should be purchased, whereas if $p < \frac{1}{3}$ the truck gardener should gamble on the weather.

In view of the preceding calculations, the truck gardener should consult the weather records to see what proportion of the past years have had a freeze during the normal growing season for lettuce. That proportion, which will be denoted by π, is the best guess, or estimate available to him for the unknown probability that a freeze will occur during the forthcoming growing season. Suppose for example, that π turned out to be .20. Then the expected profit under the two possible actions would be

$$E(I) = 24,000 \quad \text{and} \quad E(N) = 26,000.$$

This shows that the truck gardener could expect to average $2000 more by gambling on the weather than by purchasing insurance. However, for a gardener who feels for various reasons that it would be unwise to risk making only $10,000 if a freeze occurs, the principle of basing a decision on expected profit might not be altogether appealing.

3.   UTILIZING ADDITIONAL INFORMATION

In many decision making situations of the type described in the preceding section it is possible to obtain additional information concerning p by means of samples and thereby arrive at a more reliable decision. For the purpose of explaining how this is accomplished consider the following problem.

A large business firm is contemplating giving its employees some free automobile insurance as a fringe benefit. It has two plans in mind. One plan is to give each employee a $100 deductible collision insurance policy; the other is to give each employee a bonus of $30 and let him buy his own collision insurance, if he so desires. This is strictly a collision insurance policy proposal and all employees are expected to carry their own liability insurance. Although some employees with old cars would undoubtedly prefer a cash bonus to free insurance, for simplicity of exposition it will be assumed that the firm will select only one of the two plans and that after the selection has been made all employees will abide by that plan. The problem now is to determine which alternative is more economical to the firm.

Suppose that actuarial experience for the city in which the firm is located shows that for cars that have one or more accidents, the average cost per year to the insurance company for car repairs on a $100 deductible collision policy is $180. The payoff table in terms of costs to the firm per employee is shown in Table 2. The word accident in that table and in the discussion to follow will be understood to mean having an accident (at least one) to which the collision insurance is applicable and to which it is applied.

TABLE 2

		Events	
		Accident	No accident
Action	Insurance	180	0
	Bonus	30	30

Let p be the proportion of employees who will have an accident during the coming year. Then the expected costs to the firm for each employee under the two plans are

(1)
$$E(I) = 180\,p + 0(1 - p) = 180\,p$$

and

(2)
$$E(B) = 30\,p + 30(1 - p) = 30.$$

Setting $E(I) = E(B)$ and solving for p to obtain the break even point gives the value $p = \frac{1}{6}$. Consequently, if there is reason to believe that more than $\frac{1}{6}$ of the firm's employees will have an accident during the year, the bonus plan should be adopted.

Thus far this problem does not differ in principle from the problem of the preceding section. However, since this is a new experience for the firm there are no past records from which to obtain an estimate of p as was true for the other problem, which used weather records.

Suppose, instead, information is available in the form of the insurance records of all the large business firms of the city in which this particular firm is located. In particular, assume that each such firm has a record of what proportion of its employees who owned collision insurance collected on their policies during the past year. These proportions will also be represented by the letter p because they correspond to observed values of the kind of proportion involved in this problem. To simplify the computations, suppose the values of p for those firms have been classified into a relative frequency table and that they produced the distribution shown in Fig. 1.

By inspecting Fig. 1, it is clear by symmetry that the mean of the distribution is given by $\pi_0 = .175$. Since $p = \frac{1}{6}$ was found to be the break even point and the mean value of p just obtained is larger than $\frac{1}{6}$, it follows that the bonus plan should be chosen, provided the firm believes that its employees are typical

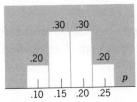

FIGURE 1. Empirical distribution of p for the business firms of a city.

insurance risks and that it is therefore willing to use $\pi_0 = .175$ as the estimate for its unknown value of p. Using π_0 for the estimate of p in this problem corresponds to using π based on weather records for p in the earlier problem, and in this sense the methods of solution are the same for the two problems.

Now if the firm in question has reason to believe that its employees are not typical risks, it can improve its decision making procedure by taking a sample of its employees and obtaining their accident experiences. Suppose a random sample of 50 employees is selected and their past year's driving records are analyzed. Suppose further that it is determined that 6 of those employees would have collected on the contemplated collision policy. Let $\hat{p} = \frac{6}{50} = .12$ denote this sample estimate of the value of p for all the employees of the firm. Since this value is smaller than $\frac{1}{6}$, the earlier decision in favor of the bonus plan is now open to question. In order to incorporate this new information into the decision making procedure, fresh values of $E(I)$ and $E(B)$ that include this information will be computed.

Let $P(p_i)$ denote the empirical probability, based on Fig. 1, that a business firm selected at random from those in that city will have $p = p_i$ for its insurance experience, and let k denote the number of such firms in the city. Then under the insurance plan the expected cost to a firm whose accident experience is p_i is given by $E_i(I) = 180\,p_i$. Now if the unknown p is treated as a random variable that can assume only the values of p listed in Fig. 1 with the corresponding probabilities given by that distribution, the expected cost to the firm in question can be obtained by calculating the expectation under that distribution assumption. It therefore follows that

$$(3) \qquad E(I) = \sum_{i=1}^{k} E_i(I)P(p_i) = \sum_{i=1}^{k} 180\,p_i P(p_i)$$

$$= 180 \sum_{i=1}^{k} p_i P(p_i) = 180\,\pi_0,$$

because the last sum is merely the mean, or expected value, of the distribution given in Fig. 1. Since the bonus plan does not involve the value of p, it is still true that

$$(4) \qquad\qquad\qquad E(B) = 30.$$

On comparing formulas (3) and (4) with formulas (1) and (2), it will be observed that these formulas merely tell one to use the mean of the distribution as the estimate of p, which was done without any formal derivation several paragraphs ago.

Although the preceding derivation did not yield a result that differs from that obtained earlier by assuming that one should use the mean of the distribution of Fig. 1 as the estimate of p, it was introduced to facilitate the use of the sampling results in constructing a more reliable decision making procedure. This will be accomplished by obtaining estimates of $P(p_i)$ that are based on both the Fig. 1 distribution and the sampling results, rather than on the distribution only.

If $P(p_i \mid S)$ denotes the conditional probability that p will assume the value p_i, given the sample result S, then it follows from Bayes' formula (10), Chapter 3, that

$$(5) \qquad P(p_i \mid S) = \frac{P(p_i)P(S \mid p_i)}{\sum\limits_{i=1}^{k} P(p_i)P(S \mid p_i)}.$$

The probabilities $P(p_i)$ obtained from Fig. 1 are called *prior probabilities* because they were obtained before the sampling experiment was conducted. The probabilities $P(p_i \mid S)$ are called *posterior probabilities,* because they are obtained after the experiment has been conducted and they contain the information S that has been supplied by the experiment.

Since the $P(p_i \mid S)$ are based on the sampling results as well as on the empirical prior distribution of p values, one would expect to obtain a more reliable estimate of $E(\mathrm{I})$ by using $P(p_i \mid S)$ in place of $P(p_i)$ in formula (3). The resulting posterior expected value, which will be denoted by $\tilde{E}(\mathrm{I})$, is therefore given by

$$\tilde{E}(\mathrm{I}) = 180 \sum_{i=1}^{k} p_i P(p_i \mid S) = 180\, \pi_1$$

where π_1 corresponds to π_0 and represents the mean of the conditional distribution. Application of formula (5) shows that

$$(6) \qquad \pi_1 = \frac{\sum\limits_{i=1}^{k} p_i P(p_i)P(S \mid p_i)}{\sum\limits_{i=1}^{k} P(p_i)P(S \mid p_i)}.$$

But if S denotes the number of successes in n trials of a binomial experiment for which p_i is the probability of success in a single trial, then

$$P(S \mid p_i) = \frac{n!}{S!(n - S)!}\, p_i{}^{S}(1 - p_i)^{n-S}.$$

Substituting this value into (6) gives

$$\pi_1 = \frac{\displaystyle\sum_{i=1}^{k} p_i P(p_i) \frac{n!}{S!(n-S)!} p_i^{S}(1-p_i)^{n-S}}{\displaystyle\sum_{i=1}^{k} P(p_i) \frac{n!}{S!(n-S)!} p_i^{S}(1-p_i)^{n-S}}.$$

Cancelling the common terms without indices reduces this to

(7)
$$\pi_1 = \frac{\displaystyle\sum_{i=1}^{k} p_i^{S+1}(1-p_i)^{n-S} P(p_i)}{\displaystyle\sum_{i=1}^{k} p_i^{S}(1-p_i)^{n-S} P(p_i)}.$$

This formula is a general formula for problems of this type in which one is given a prior distribution of p values and then performs a binomial experiment, obtaining S successes in n trials. For the purpose of applying this general formula to the problem at hand, it suffices to choose $n = 50$, $S = 6$, $k = 4$, and read off the $P(p_i)$ values from Fig. 1: $P(p_1) = .2$, $P(p_2) = .3$, $P(p_3) = .3$, $P(p_4) = .2$. Calculations carried out by means of logarithms based on these values and formula (7) will show that $\pi_1 = .145$ and hence that

$$\tilde{E}(I) = 26.$$

Since $\tilde{E}(I)$ is less than $E(B) = 30$, it now follows that the insurance plan is expected to cost less than the bonus plan. The information supplied by the sample has reversed the earlier recommended decision. Because of the low accident rate for the employees of this particular firm, it now appears more economical to give them an insurance policy.

4.   BAYESIAN METHODS

The preceding section showed how decision making could be improved by incorporating information concerning an unknown parameter p into the decision making process. There were two kinds of information used there: that obtained from knowing the distribution of p and that obtained from a sample estimate of p. Since the more information one can apply to a problem the more reliable one would expect the solution to be, the decision based on using these two kinds of information ought to be more reliable than a decision arrived at without using both of them.

Most of the statistical methods presented in the preceding chapters were based on information supplied by the sample only. Such methods are usually referred to as *classical methods*. Statistical methods that require knowledge of the probability distribution of a parameter such as p, are usually called *Bayesian methods* because of their connection with Bayes' formula. The methods employed in the preceding section were Bayesian methods because they employed the distribution of p to arrive at a solution. Although the distribution given by Fig. 1 is an empirical probability distribution, it was treated as though it were a theoretical probability distribution. The confidence that one has in a decision of the foregoing type will, of course, depend heavily on how much confidence one has that the prior distribution chosen is a realistic one.

The interpretation of probability in terms of long-run relative frequency for experiments that are repeatable, at least conceptually, implies that values arrived at from experience are estimates of underlying theoretical values. There are many situations, however, where it is unrealistic to assume that the situation being discussed is just one of similar situations to be met in the future. For example, for insurance problems accident rates vary from year to year and therefore an empirical distribution based on past experience may be out of date before it is applied. For such problems probability is thought of as a number that represents the betting odds that the event will occur. If such a probability is assigned in a problem by an individual, based on his experience or on his belief in the chances of the event occurring, it should measure the individual's willingness to give the corresponding betting odds that the event will occur. A value of p that represents an individual's degree of belief in the occurrence of an event is called *personal*, or *subjective, probability*. The latter name arises from an attempt to distinguish between probabilities whose values depend on the subject who assigns them and objective probabilities whose values are determined only by the object and the experiment in which it is involved. Thus, the probability $p = \frac{1}{6}$ that a four-spot will show in rolling a die is an objective probability because it is not based on an individual's personal judgment, whereas the probability $p = .9$ that an individual assigns to his chances of not having an automobile accident next year is a subjective probability because it is based largely on optimism and his confidence in his driving skill. There is a degree of subjectivity to all probabilities obtained from real-life experience; therefore the application of probability to real problems is bound to involve subjective probability. The principal difference in problems such as those related to games of chance and those concerned with practical situations is the degree of experience and information available for making an assessment of the pertinent probabilities.

EXERCISES

Section 2

1. Suppose you are given the option of choosing either the ice cream or hot dog concession at a baseball game. The amount of profit to be made on each of these concessions as a function of the weather is shown in the following payoff table. If the probabilities associated with the three types of weather listed are .6, .3, and .1, respectively, which concession should you choose?

		Weather		
		Sunny	Cloudy	Rain
Concession	Ice cream	300	100	−50
	Hot dog	200	200	100

2. In problem 1 delete the last column and then find what probability for sunny weather is such that the two concessions would be equally good, assuming no rain will occur.

3. A businessman is trying to decide in which of two business ventures, A or B, he should invest. He estimates the probabilities of success in the two ventures to be .2 and .4, respectively. If venture A succeeds, his profit will be $40,000. If it fails, he will lose $10,000. If venture B succeeds, his profit will be $20,000, but if it fails he will lose $12,000. Which venture should he choose?

4. A farmer must decide when to plant a crop which is sensitive to frost in its early growth stage. If he plants on March 1 and no frost occurs thereafter, he will be able to sell his early maturing crop at a profit of $10,000. If there is a frost he will lose $2000 for labor and seed. By waiting two weeks before planting, the later crop will yield a profit of $5000. If the probability of frost after March 1 is .6 and after March 15 is .2 and the farmer wishes to maximize expected profit, what date should he choose?

5. A newsboy at a newspaper stand is undecided about how many papers to order. The following table gives the probabilities (estimated) of selling the corresponding number of papers. Assume for simplicity that orders must come in units of 100 and that sales are always in units of 100. Assume also that at least 100 papers will always be sold but never more than 300. If each paper cost him 6¢ and he sells them for 10¢ and unsold papers are worthless, determine how many papers he should order to maximize expected profit.

Number	100	200	300
Probability	.3	.5	.2

6.   The purchaser of an expensive piece of electronic equipment is undecided about how many spare power tubes he should purchase when purchasing the equipment. Each spare tube will cost him $50 at the time of purchasing the unit but will cost him $150, including delay costs, if purchased later when needed for replacement. Any unused tubes are considered to be worthless. The manufacturer's records show the following experience on the needs of purchasers of this type of equipment.

(a) Construct a 4 by 4 payoff table of costs under the four possible needs and four possible purchases.

(b) Treating the table relative frequencies as probabilities, determine how many spare tubes should be purchased to minimize the expected cost of replacements.

Number of replacements	0	1	2	3
Relative frequency	.4	.3	.2	.1

7.   A record distributing company is contemplating sponsoring a record of a new singing group. It will make 50,000 records and distribute them throughout the country. Each record costs $1 to make and distribute and yields a profit of $2 if sold. Experience with similar record sponsorships shows that in 30 per cent of such sponsorships 40 per cent of the records are eventually sold, in 30 per cent of them 50 per cent of the records are sold, and in 40 per cent of them 60 per cent of the records are sold. Assuming that these constitute all the possibilities, determine whether the company should sponsor this record.

8.   In problem 7, the sales manager after listening to the record estimates that 45 per cent of the records will be sold. Assuming that his estimate is highly reliable, should the record be sponsored?

Section 3

9.   In problem 7, suppose a sample survey of several record stores is made to determine how well this particular record will sell. Letting S denote the results of the survey, assume that calculations based on the binomial distribution gave the conditional probabilities

$$P\{S|.4\} = .10, \ P\{S|.5\} = .10, \quad \text{and} \quad P\{S|.6\} = .05.$$

(a) Use these values and Bayes' formula to calculate the posterior probabilities $P\{.4|S\}$, $P\{.5|S\}$, and $P\{.6|S\}$.

(b) Using the results in (a), calculate the expected profit and determine whether the company should still sponsor the record.

CHAPTER 11

The Chi-Square Distribution

1. INTRODUCTION

In the preceding chapters problems of testing whether two means or two proportions are equal were solved by means of normal curve methods. Problems in which there are more than two variables of classification cannot, however, be solved by those methods. Suppose, for example, a business firm wishes to determine whether there are any differences in four brands of calculating machines with respect to performance time in typical business calculations. One cannot solve the problem in an efficient manner by comparing four performance time means by taking them two at a time. Or suppose a city wishes to determine whether the automobile accident rate is the same for the five working days of the week. Here also, it would be inefficient to compare only two days at a time.

In the next chapter a method will be presented for testing the equality of several means, whereas in this chapter the problem of comparing several proportions will be studied. The problem of testing the equality of two population proportions, which was solved in Chapter 8, is a special case of the problems to be considered here. Such problems can be described in the following manner.

There is a finite number, denoted by k, of possible outcomes of an experiment. These possible outcomes are represented by k cells or boxes. The experiment is performed n times, and the results are expressed by recording the observed frequencies of outcomes in the corresponding cells. The problem then is to determine whether the frequencies are compatible with those expected from some postulated theory.

As an illustration, if the experiment consists of rolling a single die, there will be 6 cells. The results of performing such an experiment sixty times are recorded in the first row of cells in Table 1. If the die is "honest," each face will have the probability $\frac{1}{6}$ of appearing in a single roll. Each face would therefore be expected to appear ten times, on the average, in an experiment of this kind. The problem then is to decide whether the observed frequencies in Table 1 are compatible with the expected frequencies listed there.

TABLE 1

	1	2	3	4	5	6
o_i	15	7	4	11	6	17
e_i	10	10	10	10	10	10

The general method for testing compatibility is based on a measure of the extent to which the observed and expected frequencies agree. This measure, called chi-square, is defined by the formula

(1)
$$\chi^2 = \sum_{i=1}^{k} \frac{(o_i - e_i)^2}{e_i}.$$

Here o_i and e_i denote the observed and expected frequencies, respectively, for the ith cell, and k denotes the number of cells.

For Table 1 the value of χ^2 is given by

$$\chi^2 = \frac{(15 - 10)^2}{10} + \frac{(7 - 10)^2}{10} + \frac{(4 - 10)^2}{10} + \frac{(11 - 10)^2}{10}$$
$$+ \frac{(6 - 10)^2}{10} + \frac{(17 - 10)^2}{10}$$

$$= 13.6.$$

Now, it is clear from inspecting formula (1) that the value of χ^2 will be 0 if there is perfect agreement with expectation, whereas its value will be large if the differences from expectation are large. Thus increasingly large values of χ^2 may be thought of as corresponding to increasingly poor experimental agreement. If an honest die were available and if the experiment of rolling the die sixty times were repeated a large number of times and each time the value of χ^2 were computed, a set of χ^2's would be obtained which could be classified into a relative frequency table and histogram of χ^2's. The histogram would tell approximately in what percentage of such experiments various ranges of values of χ^2 could be expected to be obtained. Then one would be able to judge whether the value of $\chi^2 = 13.6$ was unusually large, as compared to the run of χ^2's that are obtained in experiments with an honest die. If the percentage of experiments for which $\chi^2 > 13.6$ was very small, say less than 5 per cent, one would judge that the observed frequencies were not compatible with the frequencies expected for an honest die; hence one would conclude that the die was not honest.

2.   THE CHI-SQUARE DISTRIBUTION

Just as in the case of other sampling distributions, it is possible to use mathematical methods to arrive at the desired theoretical distribution. Since there is only a limited number of possible values for the cell frequencies in Table 1, there is only a limited number of values of χ^2 possible. Thus the theoretical distribution of χ^2 must be a discrete distribution. Since a discrete distribution with many possible values requires the application of lengthy computations, practical considerations demand a simple continuous approximation to the discrete χ^2 distribution, very much like the normal approximation to the binomial distribution. Such a continuous distribution is available in what is known as the chi-square distribution. It is unfortunate that the continuous distribution approximating the discrete chi-square distribution should also be called the chi-square distribution; however, there will be no confusion because the continuous distribution is the only one ever used.

The graph of the continuous chi-square distribution for the die problem is shown in Fig. 1. The experimental value of $\chi^2 = 13.6$ has been located on this graph, together with the value of $\chi^2 = 11.1$, which cuts off the 5 per cent right tail of the distribution. Since large values of χ^2 correspond to poor experimental agreement, the values of χ^2 exceeding 11.1 are chosen as the critical region of the test. The experimental value $\chi^2 = 13.6$ falls in the critical region; therefore, the hypothesis that the die is honest is rejected.

As a second illustration of the χ^2 test, consider the following genetics problem.

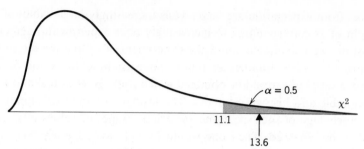

FIGURE 1.   Distribution of χ^2 for die problem.

In experiments on the breeding of flowers of a certain species, an experimenter obtained 120 magenta flowers with a green stigma, 48 magenta flowers with a red stigma, 36 red flowers with a green stigma, and 13 red flowers with a red stigma. Mendelian theory predicts that flowers of these types should be obtained in the ratios 9:3:3:1. Are these experimental results compatible with the theory?

For this problem, $n = 217$ and $k = 4$. The theoretical ratios 9:3:3:1 imply that the probabilities are $\frac{9}{16}$, $\frac{3}{16}$, $\frac{3}{16}$, and $\frac{1}{16}$ that a flower will be of the corresponding type. The expected frequencies for the four cells are therefore obtained by multiplying 217 by each of these probabilities. The observed frequencies, together with the expected frequencies, correct to the nearest integer, are shown in Table 2.

TABLE 2				
o_i	120	48	36	13
e_i	122	41	41	14

Calculations give

$$\chi^2 = \frac{(120 - 122)^2}{122} + \frac{(48 - 41)^2}{41} + \frac{(36 - 41)^2}{41} + \frac{(13 - 14)^2}{14} = 1.9.$$

Now the theory of the χ^2 distribution for problems such as these shows that the χ^2 curve in Fig. 1 does not apply to the present problem because the number of cells in this problem is not the same as before. A remarkable feature of the χ^2 distribution is that its form depends only upon the number of cells. Figure 2 gives the graphs of six such χ^2 curves corresponding to the number of cells ranging from 2 to 7. It is customary to label a χ^2 distribution by means of a parameter $\nu = k - 1$, called the number of degrees of freedom, rather than by the number of cells. The phrase degrees of freedom refers to the number of independent cell frequencies. Since the sum of the four observed frequencies in Table 2 must equal 217, the fourth cell frequency is determined as soon as the first three cell frequencies are specified. Thus there are $\nu = 3$ degrees of freedom for this problem, just as there were $\nu = 5$ degrees of freedom in the earlier problem of the die. The value of χ^2 cutting off the 5 per cent right tail of the χ^2 distribution for $\nu = 3$ turns out to be 7.8. Since the value of χ^2 for Table 2, namely 1.9, does not fall in the critical region, the result is not significant. There is no reason on the basis of this test for doubting that Mendelian theory is applicable to the data of Table 2.

The values of χ^2 that determined the 5 per cent critical regions in the two preceding illustrations were obtained from Table VII in the appendix. The 5 per cent critical value is found in the column headed .05 and in the row corresponding to the appropriate number of degrees of freedom, $\nu = k - 1$.

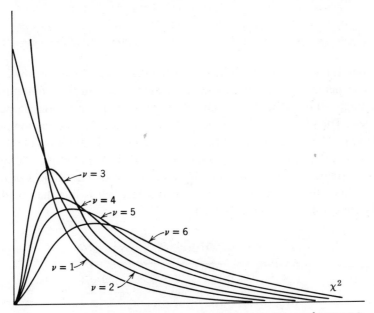

FIGURE 2.   Distribution of χ^2 for various degrees of freedom.

3.   LIMITATIONS ON THE CHI-SQUARE TEST

Since a χ^2 curve, such as those in Fig. 2, is only an approximation to the true discrete distribution of χ^2, the χ^2 test should be used only when the approximation is good. Experience and theory indicate that the approximation is usually satisfactory provided that the expected frequencies in all the cells are at least as large as 5. This limitation is similar to that placed on the use of the normal curve approximation to the binomial distribution in which np for $p \le \frac{1}{2}$ was required to exceed 5.

If the expected frequency of a cell is not as large as 5, this cell can be combined with one or more other cells until the condition is satisfied. As an illustration, consider the data of Table 3 on the classification of automobile accidents in a

TABLE 3

	15–16	17–18	19–20	21–22	23–24
o_i	5	12	10	8	5
e_i	4	8	8	10	10

certain community according to the age of the driver for drivers below 25 years of age. The percentages of drivers in these age groups in this community are approximately 10 per cent, 20 per cent, 20 per cent, 25 per cent, and 25 per cent, respectively. The problem here is to test whether the accident rate among drivers under 25 years of age is independent of age. Since the total number of observations here is $n = 40$, the expected frequencies are obtained by multiplying 40 by these percentages treated as decimals. They have been recorded in Table 3 in the row labeled e. Since the expected frequency in the first cell is less than 5, the first two cells are combined to give Table 4.

TABLE 4

15–18	19–20	21–22	23–24
17	10	8	5
12	8	10	10

The value of χ^2 for Table 4 is 5.5. Since $\nu = 3$ here and Table VII yields the 5 per cent critical value of 7.8, this result is not significant.

4.	CONTINGENCY TABLES

A very useful application of the χ^2 test occurs in connection with testing the compatibility of observed and expected frequencies in two-way tables, known as *contingency tables*. Table 5, in which the frequencies corresponding to the indicated classifications for a sample of 400 are recorded, is an illustration of a contingency table.

TABLE 5

Education	Marriage Adjustment				Totals
	Very Low	Low	High	Very High	
College	18(27)	29(39)	70(64)	115(102)	232
High school	17(13)	28(19)	30(32)	41(51)	116
Grade school	11(6)	10(9)	11(14)	20(23)	52
Totals	46	67	111	176	400

A contingency table is usually constructed for the purpose of studying the relationship between the two variables of classification. In particular, one may wish to know whether the two variables are at all related. By means of the χ^2 test it is possible to test the hypothesis that the two variables are independent. Thus, in connection with Table 5, the χ^2 test can be used to test the hypothesis that there is no relationship between an individual's educational level and his adjustment to marriage.

This problem differs from the preceding problems in that the probabilities of an observation falling in the various cells are not known. As a result, it is not possible to write down the expected frequencies for the various cells, as was the case in the other problems. This difficulty can be overcome in the following manner.

Consider repeated sampling experiments of this kind in which 400 people are classified in their proper categories. If only those experiments that produce the same marginal totals are considered, then expected frequencies can be obtained.

Since the margins are now fixed, the proportion of college graduates in such samples of 400 is always $232/400 = .58$. Therefore, if there is no relationship between education and marriage adjustment, 58 per cent of the 46 individuals in the very low adjustment category would be expected to be college graduates. Since 58 per cent of 46 is 27, to the nearest unit, this is the expected frequency for the first cell in Table 5 on the basis of independence. The expected frequencies in the remaining three cells of the first row are obtained in a similar manner by taking 58 per cent of the column totals. The second- and third-row cell frequencies are obtained by using 29 per cent and 13 per cent, respectively, corresponding to the proportions $116/400$ and $52/400$. The expected frequencies are the frequencies recorded in parentheses in Table 5.

The value of χ^2 for Table 5, using the numbers in parentheses as expected frequencies, is 20.7. Although there are twelve cells in this table, one does not choose $\nu = 11$ degrees of freedom in looking up the critical value of χ^2, as was the procedure in earlier problems. The correct value of ν to choose here is quite different and is determined by strictly mathematical arguments; however, it is possible to acquire a feeling for the plausibility of the correct value by the following type of argument. Assume that n is large but fixed in value. Then χ^2 will be expected to decrease in variability if the number of cells is decreased, because there will be fewer possibilities for large differences between observed and expected frequencies. This is observed in Fig. 2 in the shifting of the distribution to the left as ν decreases. Furthermore, if the number of cells is fixed but restrictions are placed on the frequencies that occur in the cells, the variability of χ^2 will also be expected to decrease. Now in repeated experiments of the preceding type the cell frequencies have been rather severely restricted by requiring them to possess the proper row sums and the proper column sums. In the first row of Table 5, for example, the observed frequencies must sum to 232. This essentially says that the frequencies in the first three cells of that row are free to assume any values they please, as long as they do not sum to more than 232, but then the fourth cell frequency is completely determined. Thus, if the χ^2 test were to be applied to the first row only it would employ $\nu = 3$ degrees of freedom. The phrase "degrees of freedom" aptly describes what corresponds to the number of cell frequencies that are free to vary. The same argument would apply to the second row, thus giving a total of six degrees of freedom for the two rows. However, when one looks at the third row, the picture changes because now it is necessary to realize that what applies to rows must also apply to columns; therefore the frequency found in the last cell of the first column is determined when the first two cell frequencies in that column have been specified. As a result, the third-row frequencies as well as the fourth-column frequencies are determined when the remaining cell frequencies have been specified. Thus, it seems reasonable that the experiment should behave like one in which there are six cells in which the frequencies are free to vary, subject only to the restriction that their sums are not excessive. In view of arguments of this type it should come as no surprise to learn that mathematical theory has demonstrated that one should choose $\nu = 6$ here.

From Table VII, the 5 per cent critical value of χ^2 with $\nu = 6$ degrees of freedom is 12.6. Since $\chi^2 = 20.7$ here, this result is significant and the hypothesis of independence is rejected. An inspection of Table 5 shows that individuals with some college education appear to adjust themselves to marriage more readily than those with less education.

In the foregoing solution, since only experiments with the same marginal totals are being considered, it is tacitly assumed that any relationship, or lack of it, that exists for restricted sampling experiments will also hold for unrestricted sampling experiments. This seems to be a reasonable assumption here because there appears to be no reason for believing that fixing the marginal totals in this way will influence the relationship.

This problem illustrates very well the reason why Table VII for χ^2 lists the number of degrees of freedom ν rather than the number of cells to determine which χ^2 curve to use. For a contingency table having r rows and c columns, the number of degrees of freedom is given by the formula

$$\nu = (r - 1)(c - 1).$$

This follows from the earlier arguments that the frequencies in the last row and in the last column are determined by the marginal totals as soon as the other cell frequencies are given. Thus the number of independent cell frequencies is obtained by counting the number of cells after the last row and the last column have been deleted. After the deletion there will be $r - 1$ rows and $c - 1$ columns and therefore $(r - 1)(c - 1)$ cells.

Although the distribution of χ^2 depends only on the number of cells in a contingency table, the value of χ^2 tends to grow as the sample size increases even though the row and column proportions remain the same. To observe the effect of increasing the sample size, consider the following problem.

A sample of 120 students was classified with respect to appearance and with respect to grade point average. Three categories of appearance were used: attractive, ordinary, and unattractive, and four categories of grade point averages were used. The latter were chosen from high to low to give approximately the same number of students in each category. This classification produced the following contingency table values shown in Table 6. The problem is to test the hypothesis that there is no relationship between the two variables of classification.

TABLE 6

	H	H-M	M-L	L
A	14	11	10	5
O	10	16	16	14
U	3	4	7	10

Summing the frequencies in rows and columns gives the marginal totals shown in Table 7. Taking the proportion $\frac{40}{120}$, or $\frac{1}{3}$, of each column total gives the expected frequencies, in parentheses, in the first row. The expected frequencies for the second and third rows were obtained in a similar manner by taking the proportions $\frac{56}{120}$ and $\frac{24}{120}$ of the column totals. The calculations here were to the nearest integer only.

TABLE 7

14 (9)	11 (10)	10 (11)	5 (10)	40
10 (13)	16 (14)	16 (15)	14 (14)	56
3 (5)	4 (6)	7 (7)	10 (6)	24
27	31	33	29	120

The calculation of χ^2 then proceeds as follows:

$$\chi^2 = \frac{(14-9)^2}{9} + \frac{(11-10)^2}{10} + \frac{(10-11)^2}{11} + \frac{(5-10)^2}{10}$$
$$+ \frac{(10-13)^2}{13} + \frac{(16-14)^2}{14} + \frac{(16-15)^2}{15} + \frac{(14-14)^2}{14}$$
$$+ \frac{(3-5)^2}{5} + \frac{(4-6)^2}{6} + \frac{(7-7)^2}{7} + \frac{(10-6)^2}{6} = 10.7.$$

Since $\chi_0^2 = 12.59$ and $\chi^2 < 12.59$, accept the hypothesis of no relationship.

Now suppose each frequency is doubled. Then the expected frequencies will also be doubled. Each term in χ^2 will therefore be multiplied by 2. This is seen by looking at the first term which will become

$$\frac{(28 - 18)^2}{18} = \frac{[2(14 - 9)]^2}{2(9)} = 4\frac{(14 - 9)^2}{2(9)} = 2\frac{(14 - 9)^2}{9}.$$

As a result $\chi^2 = 21.4$. Since the degrees of freedom have not changed, $\chi_0^2 = 12.59$. Now the hypothesis is rejected. Thus, even though the marginal proportions remain the same, the value of χ^2 tends to grow as the same size increases, provided the two variables are not independent. Thus, the larger the value of n, the more easily is a lack of independence discovered.

EXERCISES

Section 2

1. According to Mendelian inheritance, offspring of a certain crossing should be colored red, black, or white in the ratios $9:3:4$. If an experiment gave 72, 32, and 40 offspring in those categories, is the theory substantiated?

2. The number of individuals of a certain race possessing the four blood types should be in the proportions .16, .48, .20, .16. Given the observed frequencies 180, 360, 130, 100 for another race, test to see whether it possesses the same distribution of blood types.

3. In a certain city the proportion of car owners who have 0 accidents, 1 accident, and more than 1 accident in a year are, respectively, .75, .20, and .05. An insurance company took a sample of 100 of its insured and found 66, 22, and 12 in those categories. Are these frequencies compatible with the postulated theoretical proportions?

4. In the preceding problem combine the frequencies in the last two cells and test. Comment on any differences in the two tests.

5. A sample of size 100 was taken from the distribution given by Fig. 5, Chapter 6, and produced the following frequencies corresponding to the six possible values of x: 22, 28, 17, 18, 7, 8. These frequencies were obtained by the use of Table II in the manner indicated in connection with the sampling experiment of that chapter. Test to see whether this sample is compatible with the probabilities of Fig. 5, Chapter 6.

6. The number of automobile accidents per week in a certain community were as follows: 12, 7, 20, 2, 14, 9, 16, 6, 10, 4. Are these frequencies in agreement with the belief that accident conditions were the same over this 10-week period?

7. Take a sample of 500 random digits from Table II and list the observed frequencies in 10 cells. (a) Apply the χ^2 test to see whether the sample is compatible with theory here. (b) Combine the odd-digit cells and the even-digit cells to obtain only two cells and test to see whether odd and even digits possess the same probability of occurrence. (c) Could you have worked part (b) by an earlier technique? If you can, do so and compare your two results.

Section 3

8. A pinball machine has six holes of varying size into which the ball may land. The payoff for the ball landing in a given hole indicates that the probabilities of success for the various holes should be $\frac{6}{21}, \frac{5}{21}, \frac{4}{21}, \frac{3}{21}, \frac{2}{21}$, and $\frac{1}{21}$. A contestant plays the game 42 times and obtains the following frequencies: 19, 10, 6, 5, 2, and 0. Test the hypothesis that the probabilities corresponding to the payoffs are correct.

9. Suppose one wishes to test the hypothesis that the payoffs for the more difficult holes are too low as compared to the payoffs for the easier holes in the game of problem 8. One method of doing this is to combine the frequencies of the first two holes and of the last four holes and test the hypothesis $H_0 : p = \frac{11}{21}$, where p is the probability of landing in one of the first two holes. Carry out this test.

10. The following data represent the results of an investigation of the sex distribution of the children of 32 families containing 4 children each. Use the binomial distribution with $n = 4$ and $p = \frac{1}{2}$ to calculate expected frequencies. Then apply the χ^2 test to see whether this binomial distribution model is satisfactory here.

No. of sons	0	1	2	3	4
No. of families	4	10	8	7	3

11. Test to see whether the two variables of classification in the following contingency table are independent.

20	10	10
10	20	30

12. A certain drug is claimed to be effective in curing colds. In an experiment on 164 people with colds, half of them were given the drug and half of them were given sugar pills. The patients' reactions to the treatment are recorded in the following table. Test the hypothesis that the drug and the sugar pills yield similar reactions.

	Helped	Harmed	No Effect
Drug	50	10	22
Sugar	44	12	26

13. A market analyst is concerned whether housewives who are not at home when interviewers call differ in their opinions of a certain product. To check this possibility, interviewers returned to "not-at-home" houses until an interview was obtained. The results of this study are given in the following table. Test to see whether the "not-at-home" housewives have the same opinion as the "at-home" housewives.

Opinion of Product	Number of Housewives Interviewed	
	First Call	Later Call
Excellent	62	34
Satisfactory	84	44
Unsatisfactory	24	22

14. In an epidemic of a certain disease 927 children contracted the disease. Of these, 408 received no treatment and 104 of them suffered aftereffects. Of the remainder who did receive treatment, 166 suffered aftereffects. Test the hypothesis that the treatment was not effective and comment about the conclusion.

15. Work problem 14 by the method explained in Chapter 8 for testing the difference of two proportions. It can be shown that the two methods are equivalent for problems such as this.

16. The following table is based on a survey of doctors in a certain city. They were divided into three groups on the basis of their academic record in medical school and their income in practice fifteen years after graduation. Test to see whether there is any relationship between these two variables.

		Income		
		High	Medium	Low
Academic Record	High	18	17	5
	Medium	26	38	16
	Low	6	15	9

$$V = k - 1$$
$$V = (r-1)(c-1)$$

CHAPTER 12

Analysis of Variance

Perhaps the problem that arises more frequently than any other in statistical work is that of testing whether two samples differ significantly with respect to some property. In Chapter 8 this problem was solved by testing the equality of two population means or two population proportions.

The reason this type of problem occurs so frequently is that experimentalists often design an experiment to compare a new technique or process with a standard one. For example, an educator may believe that he has discovered a better way of teaching foreign languages than that being used at his institution; or a chemist may have discovered a new plastic that he believes will be superior to the one being manufactured at his plant. In either case, an experiment would be designed to test whether the new method or material was in fact superior to the old one.

As was pointed out in the preceding chapter, situations often occur, however, in which there are several methods or products rather than just two that are competing against each other. For example, a manufacturer of cake mixes may vary the amount of a certain ingredient to obtain six different mixes that he would like to compare for quality. Now it is usually very inefficient to compare several samples by comparing them two at a time. If one had, say, six samples to compare, there would be fifteen such comparison pairs. Furthermore, the probability associated with testing a single difference is no longer applicable when testing several differences simultaneously. Another disadvantage of the method based upon comparing only pairs of samples is that experimentalists who are accustomed to comparing only two samples at a time may be led into designing poor experiments to accomplish their ultimate objective. The manufacturer of cake mixes, for example, who changed only a single ingredient at a time and then retained only the better of two mixes each time might well miss out on a much better mix obtained by varying several ingredients by different amounts and considering various combinations of mixes simultaneously. In agricultural experiments concerned with testing different types and amounts of fertilizers and different seed varieties it has been found very inefficient not to consider different combinations of those quantities simultaneously.

In view of the foregoing discussion, it seems clear that a new method is needed to solve some of the problems related to several sample means. One of the methods that has been designed to solve problems of this type is known as the *analysis of variance*. As the name might indicate, the method consists of analyzing the variance of the sample into useful components. Although the method has been developed to treat a wide variety of problems, only two applications are considered in this chapter.

2.   ONE VARIABLE OF CLASSIFICATION

The simplest type of analysis of variance model is the one in which observations are classified into groups on the basis of a single property. For example, in studying the marketing weights of rabbits, one might wish to classify rabbits on the basis of the number in a litter; or, in studying the degree of political conservatism in voters, one might wish to classify voters on the basis of income.

For the purpose of explaining the analysis of variance technique, consider the data of Table 1. These data represent the scores made by 24 soldiers in an experiment to determine whether shooting accuracy is affected by methods of sighting: (a) with the right eye open, (b) with the left eye open, (c) with both eyes open. Twenty-four soldiers were selected at random from a certain training base and then split into three equal groups corresponding to the three sighting

TABLE 1

Right	Left	Both
44	40	51
39	37	47
33	28	37
56	53	52
43	38	42
56	51	63
47	45	46
58	60	62

methods. After each group had become familiar with its sighting method, the experiment was conducted by allowing each participant to shoot the same number of rounds at a target.

If there is no advantage in one sighting method over either of the others, these scores could be treated as those of 24 randomly selected soldiers who used any one of the three sighting methods. It is assumed that such scores are normally distributed. However, if the three sighting methods are not equally good, it is necessary to modify the normality assumption to the extent of assuming that each of the three sighting methods has its own normal distribution of scores. In problems like this, when differences occur they usually occur in the means rather than in the standard deviations; therefore, it is assumed that the standard deviations of the three normal populations are equal.

If the population means corresponding to the three sighting methods are denoted by $\mu_1, \mu_2,$ and μ_3, then the problem reduces to one of testing the hypothesis

$$H_0 : \mu_1 = \mu_2 = \mu_3.$$

The foregoing assumptions are the same as those made in Chapter 7 for testing the equality of two means by small sample methods and which permitted the use of Student's t distribution. Unfortunately that distribution cannot be used in solving the present problem and therefore a new approach is needed. Such an approach can be based on comparing two different estimates of the common variance, σ^2, of the three populations.

If the hypothesis H_0 is true the classification of the data into three columns is meaningless and the entire set of measurements can be treated as a sample of size 24 from a normal population. If σ^2 denotes the variance of this population, an estimate of σ^2 can be obtained by means of the familiar sample variance based on those 24 measurements. However, there are several other ways of obtaining valid estimates of σ^2. For example, the sample variance of the first column of measurements is a valid estimate of σ^2 although it is not nearly as good as the estimate based on all the measurements. Similarly, the sample variances of the second and third columns are also valid estimates of σ^2. Furthermore, the mean of the three column estimates is a valid estimate of σ^2 and nearly as good as the familiar estimate based on combining the three sets of measurements. If s_1^2, s_2^2, and s_3^2 denote the sample variances for the three columns, this last estimate, which will be denoted by V_c, is

(1)
$$V_c = \frac{s_1^2 + s_2^2 + s_3^2}{3}.$$

The subscript c is used here to indicate that the estimate is based on the column variances.

Another quite different type of estimate of σ^2 can be obtained by using the relationship between the variance of a sample mean and the variance of the population, namely $\sigma_{\bar{x}}^2 = \sigma^2/n$. It is convenient here to express this relationship in the form

(2)
$$\sigma^2 = n\sigma_{\bar{x}}^2$$

Suppose several samples of size n each have been taken from some population. If the sample means have been calculated, then the sample variance of those sample means will be a valid estimate of $\sigma_{\bar{x}}^2$. In general, the sample variance of a set of measurements is a valid estimate of the population variance of the measurements regardless of whether those measurements happen to be simple measurements, or means of simple measurements, or other functions of simple measurements. From (2) it follows that if an estimate of $\sigma_{\bar{x}}^2$ is available, it may be multiplied by n to yield an estimate of σ^2. In the present problem there are three such sample means that may be used to construct an estimate of $\sigma_{\bar{x}}^2$. They are the three column means, which will be denoted by $\bar{x}_1$, $\bar{x}_2$, and $\bar{x}_3$. If $\tilde{x}$ denotes the mean of those three column means, then an estimate of $\sigma_{\bar{x}}^2$ is given by $\sum_{i=1}^{3} (\bar{x}_i - \tilde{x})^2/2$. Since those means are based on samples of size 8 each, it follows that $n = 8$ and hence from (2) that the desired estimate of σ^2 is given by

(3)
$$V_m = 8\frac{\sum\limits_{i=1}^{3} (\bar{x}_i - \tilde{x})^2}{2}.$$

The subscript m is used here to indicate that the estimate is based on the means of the columns.

Since V_c and V_m are both valid estimates of σ^2 when H_0 is true, it follows that they should be approximately equal in value, and therefore that their ratio should have a value close to 1. If, however, H_0 is not true and the column means differ considerably, the two estimates V_c and V_m will be seen to differ considerably in value. Because the estimate V_c is based on calculating the variances of each column separately, it will be unaffected by changing the means of the various columns, for the variance of a set of measurements is independent of the value of their mean. It is clear from (3), however, that the estimate V_m will be directly affected and will increase in value as the sample means move apart. Thus it appears that the ratio of V_m and V_c will differ considerably from 1 when H_0 is not true. This ratio will be used as the desired quantity for testing the hypothesis H_0. It is denoted by the letter F; hence

$$(4) \qquad\qquad\qquad F = \frac{V_m}{V_c}.$$

The discussion thus far has been mostly of the qualitative type, stating that F can be expected to have a value close to 1 when H_0 is true and to have a value considerably larger than 1 when H_0 is not true and the population means differ widely. This information is not sufficient for constructing a test based on probability; it is necessary to know what the distribution of F is before such a test can be performed.

Just as in the case of other sampling distributions, it is possible to approximate the distribution of F by carrying out repeated sampling experiments of the type being considered here and constructing the histogram of the resulting F values; however, the exact sampling distribution of F can be obtained by mathematical methods. It turns out that under the normality assumption the distribution of F depends only upon how many data were available for the numerator estimate of σ^2 and how many were available for the denominator estimate. Table IX in the appendix lists the 5 per cent and the 1 per cent right tail critical values of F corresponding to different values of the parameters ν_1 and ν_2, which are called the number of degrees of freedom in the numerator and denominator of F.

The degrees of freedom here are those that one would naturally associate with the sample variances being used. Since the number of degrees of freedom for the usual estimate of σ^2 is given by $\nu = n - 1$, or one less than the number of measurements, the number of degrees of freedom for the numerator of F in this problem is given by $\nu_1 = 2$ because the estimate is based on the three sample means. The number of degrees of freedom for the denominator of F in this problem is $\nu_2 = 21$ because each column variance contributes 7 degrees of freedom and all three column variances are employed.

Now return to the particular problem that motivated the preceding discussion. Calculations with the data of Table 1 yield the values $\bar{x}_1 = 47$, $\bar{x}_2 = 44$, $\bar{x}_3 =$

50, $s_1^2 = 81.1$, $s_2^2 = 106.3$, and $s_3^2 = 82.3$. As a result, it follows from (1) that $V_c = 89.9$. Additional calculations yield the value $\Sigma(\bar{x}_i - \check{x})^2/2 = 9$; hence it follows from (3) that $V_m = 72$. The value of F is therefore given by

$$F = \frac{V_m}{V_c} = \frac{72}{89.9} = .80.$$

From Table IX it will be found that the 5 per cent critical value of F corresponding to $v_1 = 2$ and $v_2 = 21$ is 3.47. Since $F = .80$ for this problem, the hypothesis is accepted. The data are in agreement with the view that accuracy of shooting is not affected by which of the three sighting methods is used. It would not have been necessary to consult Table IX for this problem because F values in the neighborhood of 1 are to be expected when H_0 is true, and therefore a value of $F < 1$ could not possibly lie in the critical region of large F values.

Although the approach used to arrive at the F variable for testing the hypothesis that a set of column means are equal seems quite different from that used in Chapter 8 for testing the equality of two column means, it can be shown that the F test, when applied to testing the equality of two column means, is equivalent to the t test for the same problem. Thus, the test based on F is a generalization of the earlier two-column test based on t.

The foregoing problem is a special case of more general problems of this type in which the number of sample values in each column is not necessarily the same. For the purpose of obtaining a formula that will treat these more general problems it is convenient to introduce the following notation.

Let x_{ij} denote the ith measurement in the jth column of a table of measurements of which Table 1 is a special case. Let c denote the number of columns in this table, and let n_j denote the number of measurements in the jth column. This notation is displayed in Table 2.

The mean of the jth column measurements is denoted by the symbol $\bar{x}_{\cdot j}$. The dot is placed in front of the j to indicate that the mean was obtained from summing on the index i (rows). In section 3 it is necessary to sum over columns as well; therefore, some notation such as this is needed to keep straight whether one is summing over rows or columns.

In terms of this new notation, the numerator of F is given by the formula

(5)
$$V_m = \frac{\sum_{j=1}^{c} n_j(\bar{x}_{\cdot j} - \bar{x})^2}{c - 1}$$

Here $\bar{x}$ denotes the mean of the entire set of measurements. It replaces $\check{x}$, which was used earlier in (3). If the n_j are all equal, and therefore equal to the number

TABLE 2

$$
\begin{array}{ccccc}
x_{11} & x_{12} & \cdots & x_{1j} & \cdots & x_{1c} \\
x_{21} & x_{22} & \cdots & x_{2j} & \cdots & x_{2c} \\
\vdots & \vdots & & \vdots & & \vdots \\
x_{i1} & x_{i2} & \cdots & x_{ij} & \cdots & x_{ic} \\
\vdots & & & \vdots & & \vdots \\
x_{n_11} & \vdots & & x_{nj} & \\
 & x_{n_22} & & & & x_{n_cc} \\
\hline
\bar{x}_{\cdot1} & \bar{x}_{\cdot2} & \cdots & \bar{x}_{\cdot j} & \cdots & \bar{x}_{\cdot c}
\end{array}
$$

of rows in the table, V_m will have the same form as that in (3) because in that problem there were eight rows and three columns.

In terms of this same notation, the denominator of F is given by the formula

(6)
$$
V_c = \frac{\displaystyle\sum_{j=1}^{c}\sum_{i=1}^{n_j}(x_{ij} - \bar{x}_{\cdot j})^2}{\displaystyle\sum_{j=1}^{c}(n_j - 1)}.
$$

If the n_j are all equal this will have the same form as that in (1) because the following quantity, which is denoted by s_j^2,

$$
s_j^2 = \frac{\displaystyle\sum_{i=1}^{n_j}(x_{ij} - \bar{x}_{\cdot j})^2}{n_j - 1}
$$

is the sample variance of the jth column measurements and therefore (6) reduces to

$$
V_c = \frac{s_1^2 + s_2^2 + \cdots + s_c^2}{c}.
$$

The ratio of (5) and (6) supplies the desired F variable; with the degrees of freedom given by $\nu_1 = c - 1$ and $\nu_2 = \displaystyle\sum_{j=1}^{c}(n_j - 1)$; hence

(7)
$$
F = \frac{\displaystyle\sum_{j=1}^{c} n_j(\bar{x}_{\cdot j} - \bar{x})^2}{\displaystyle\sum_{j=1}^{c}\sum_{i=1}^{n_j}(x_{ij} - \bar{x}_{\cdot j})^2} \cdot \frac{\displaystyle\sum_{j=1}^{c}(n_j - 1)}{c - 1}
$$

3.   TWO VARIABLES OF CLASSIFICATION

The foregoing analysis of variance problem was relatively simple because there was only one classification variable, namely the method of sighting. In an experiment of this type one could consider many other variables that might influence the accuracy of shooting. For example, one might compare different positions of firing, different brands of guns, or different brands of bullets. Analysis of variance methods have been designed to treat any number of classification variables; however, the discussion here is limited to two such variables. The methods for more than two variables are very similar to those for two variables.

The problem considered in section 2 can be modified to yield a problem that would normally be solved by the methods of this section. Assume that instead of having drawn a random sample of 24 soldiers from the base a random sample of only 8 soldiers had been drawn. Then it would have been necessary for each of the 8 soldiers to become acquainted with all three methods of sighting. After that, each soldier would have been instructed to fire the same number of rounds with each of the three sighting methods. Thus the three scores in the first row of Table 1 would represent the three scores made by the first soldier. In such an experiment one would randomize the order in which the different sighting methods were employed by the marksmen so that none of the methods would have an advantage with respect to practice.

The second variable of classification here is the individual. Since it is well known that there is large variation in the skill of individuals in shooting, it would seem desirable to control this feature of the variability of scores so that any differences that might be caused by the different sighting methods could be recognized. Large individual differences among the 24 soldiers of the earlier experiment might obliterate any moderate differences arising because of the different sighting methods.

Since there will now be the same number of measurements in each column, Table 2 will assume a neater appearance. Denoting the number of rows by r, it will assume the form shown in Table 3. In this table the row means are also displayed because they will be needed presently.

In the two variable analysis of variance setup one assumes that each of the variables x_{ij} in Table 3 is an independent normal variable with a common variance σ^2. This means, for example, that in repeated gunnery experiments of the type being discussed the same 8 soldiers would use each of the three sighting methods each time the experiment was run. This differs from the one variable of classification setup in which fresh sets of 24 soldiers would be selected each time. For

TABLE 3

$$
\begin{array}{ccccc|c}
x_{11} & x_{12} & \cdots & x_{1j} & \cdots & x_{1c} & \bar{x}_{1\cdot} \\
x_{21} & x_{22} & \cdots & x_{2j} & \cdots & x_{2c} & \bar{x}_{2\cdot} \\
\vdots & \vdots & & \vdots & & \vdots & \vdots \\
x_{i1} & x_{i2} & \cdots & x_{ij} & \cdots & x_{ic} & \bar{x}_{i\cdot} \\
\vdots & \vdots & & \vdots & & \vdots & \vdots \\
x_{r1} & x_{r2} & \cdots & x_{rj} & \cdots & x_{rc} & \bar{x}_{r\cdot} \\
\hline
\bar{x}_{\cdot 1} & \bar{x}_{\cdot 2} & \cdots & \bar{x}_{\cdot j} & \cdots & \bar{x}_{\cdot c} & \bar{x}
\end{array}
$$

the two variable situation it is also necessary to make a few more assumptions about the basic variables, but these assumptions are not discussed here.

Just as in the earlier method, one finds two estimates of the common variance σ^2 and then uses the ratio of the two estimates to obtain an F value. The method of finding such estimates is based upon taking the natural variance estimate and analyzing it into useful components. This procedure gave rise to the name "analysis of variance." The natural variance estimate for Table 3 is the quantity

$$\frac{\displaystyle\sum_{i=1}^{r}\sum_{j=1}^{c}(x_{ij}-\bar{x})^2}{rc-1}.$$

Only the numerator of this estimate is used in the following analysis. Now it can be shown by simple algebraic manipulations that the following formula holds:

$$(8) \quad \sum_{i=1}^{r}\sum_{j=1}^{c}(x_{ij}-\bar{x})^2 = \sum_{i=1}^{r}\sum_{j=1}^{c}(\bar{x}_{i\cdot}-\bar{x})^2 + \sum_{i=1}^{r}\sum_{j=1}^{c}(\bar{x}_{\cdot j}-\bar{x})^2$$

$$+ \sum_{i=1}^{r}\sum_{j=1}^{c}(x_{ij}-\bar{x}_{i\cdot}-\bar{x}_{\cdot j}+\bar{x})^2.$$

It can also be shown that each of the three sums of squares on the right side of (8), if divided by the proper constant, is a valid estimate of σ^2 when it is assumed that there are no real differences in the row means or the column means. These three estimates are

$$V_r = \frac{\displaystyle\sum_{i=1}^{r}\sum_{j=1}^{c}(\bar{x}_{i\cdot}-\bar{x})^2}{r-1} = c\frac{\displaystyle\sum_{i=1}^{r}(\bar{x}_{i\cdot}-\bar{x})^2}{r-1},$$

$$V_c = \frac{\sum_{i=1}^{r} \sum_{j=1}^{c} (\bar{x}_{\cdot j} - \bar{x})^2}{c - 1} = r\frac{\sum_{j=1}^{c} (\bar{x}_{\cdot j} - \bar{x})^2}{c - 1},$$

$$V_e = \frac{\sum_{i=1}^{r} \sum_{j=1}^{c} (x_{ij} - \bar{x}_{i\cdot} - \bar{x}_{\cdot j} + \bar{x})^2}{(r - 1)(c - 1)}.$$

The subscripts here on V refer to rows, columns, and experimental error. The expression for V_c is the same as that for V_m given by (5) when $n_j = r$. The expression for V_r is similar, except that it measures row variation rather than column variation; therefore, if there are large differences in shooting skill among soldiers, this quantity will tend to be considerably larger than if there were no individual differences. The expression for V_e essentially measures the variation in the data after the variation caused by column differences and row differences has been eliminated. It serves as an estimate of σ^2 unaffected by sighting differences and individual differences.

For the purpose of testing the hypothesis that all the theoretical column means are equal, the estimates to use are V_c and V_e. Thus the test reduces to computing the value of F given by

(9)
$$F = \frac{r(r - 1) \sum_{j=1}^{c} (\bar{x}_{\cdot j} - \bar{x})^2}{\sum_{i=1}^{r} \sum_{j=1}^{c} (x_{ij} - \bar{x}_{i\cdot} - \bar{x}_{\cdot j} + \bar{x})^2},$$
$$\nu_1 = c - 1, \qquad \nu_2 = (r - 1)(c - 1).$$

The values of ν_1 and ν_2 in F are always the denominators needed to make the corresponding sums of squares valid estimates of σ^2. The mathematical theory for this problem shows that the F distribution is valid here whether or not there are real differences in the row means.

In the two variable scheme it is also possible to test the hypothesis that the theoretical row means are equal, which implies that there are no differences among the 8 soldiers with respect to shooting skill. Here one uses the estimates V_r and V_e, and one forms the F ratio given by

(10)
$$F = \frac{c(c - 1) \sum_{i=1}^{r} (\bar{x}_{i\cdot} - \bar{x})^2}{\sum_{i=1}^{r} \sum_{j=1}^{c} (x_{ij} - \bar{x}_{i\cdot} - \bar{x}_{\cdot j} + \bar{x})^2},$$
$$\nu_1 = r - 1, \qquad \nu_2 = (r - 1)(c - 1).$$

In view of the fact that the more sources of variation one can control in an experiment the more likely one is to detect differences of experimental interest, the F test based on eliminating variation due to individual differences in shooting skill and given by formula (9) should be a more delicate test than the one used in section 2 and given by formula (7). For the purpose of comparing these two tests, consider the application of formula (9) to the data of Table 1.

The numerator sum of squares is the same as in the earlier test; therefore, its numerical value need not be computed. It is usually easier to compute the denominator sum of squares by means of formula (8) than it is to compute it directly from its definition. Thus one computes the left side of (8) as well as the first two sums of squares on the right side and then obtains the desired sum of squares by subtraction. Earlier computations gave

$$\sum_{i=1}^{8} \sum_{j=1}^{3} (\bar{x}_{\cdot j} - \bar{x})^2 = 8(18) = 144.$$

The first sum of squares on the right side of (8) was computed in the same manner as the second sum of squares. Computations for the data of Table 1 yielded the values

$$\sum_{i=1}^{8} \sum_{j=1}^{3} (x_{ij} - \bar{x})^2 = 2032,$$

and

$$\sum_{j=1}^{3} \sum_{i=1}^{8} (\bar{x}_{i\cdot} - \bar{x})^2 = 1768.$$

Consequently, formula (8) yields the value

$$\sum_{i=1}^{8} \sum_{j=1}^{3} (x_{ij} - \bar{x}_{i\cdot} - \bar{x}_{\cdot j} + \bar{x})^2 = 2032 - 1768 - 144 = 120.$$

The value of F as given by formula (9) then becomes

$$F = \frac{8 \cdot 7(18)}{120} = 8.4,$$

$$\nu_1 = 2, \qquad \nu_2 = 14.$$

From Table IX it will be found that the 5 per cent critical value of F corresponding to $\nu_1 = 2$ and $\nu_2 = 14$ is 3.74. Since $F = 8.4$ is in the critical region, the hypothesis H_0 is rejected.

The conclusion here is contrary to that made for the same data in section 2. Actually, the data for Table 1 were obtained for a group of 8 soldiers; consequently, only the second method is applicable here. The first method required that the scores should be those for 24 randomly selected soldiers. The purpose of using the same data for both methods was to point out the similarities and the differences of the two methods and to stress the fact that it usually pays to introduce important classification variables in the analysis of variance technique if one wishes to obtain a delicate test for testing a set of theoretical means. If there had been no appreciable individual differences in shooting skill, nothing would have been gained by designing the experiment to measure and eliminate this source of variation in the test, in which case the method in section 2 would have been preferable. The application of formula (10) shows, however, that $F = 29.5$, $v_1 = 7$, $v_2 = 14$. Since the 5 per cent critical value of F is 2.77, this means that there is large variation in the row means, hence large variation in individual shooting skill.

The analysis of variance method of analyzing experimental data received much of its stimulus and development from statisticians who worked at agricultural experiment stations. They found it to be a highly efficient method of designing and analyzing experiments in agriculture. The following example, which is concerned with determining whether there are any essential differences in five different fertilizers, is a simple illustration of their type of problems.

Four plots of ground were available for the experiment. Each plot was divided into five subplots and each of the five fertilizers was assigned at random to one of these subplots. This was done for each of the four plots of ground. Potatoes were grown on the plots and fertilized according to the preceding assignments. At the end of the growing season the yield of potatoes on each subplot was obtained and produced the data shown in Table 4. The problem is to test whether the five fertilizers are equally effective with respect to the mean yield of potatoes.

TABLE 4

Plot	Fertilizer				
---	A	B	C	D	E
1	310	353	366	299	367
2	284	293	335	264	314
3	307	306	339	311	377
4	267	308	312	266	342

Calculations give

$$\bar{x}_{.1} = 292, \ \bar{x}_{.2} = 315, \ \bar{x}_{.3} = 338, \ \bar{x}_{.4} = 285, \ \bar{x}_{.5} = 350,$$
$$\bar{x}_{1.} = 339, \ \bar{x}_{2.} = 298, \ \bar{x}_{3.} = 328, \ \bar{x}_{4.} = 299, \ \bar{x} = 316.$$

Additional calculations give

$$\sum_{i=1}^{4} (\bar{x}_{i.} - \bar{x})^2 = 1286, \ \sum_{j=1}^{5} (\bar{x}_{.j} - \bar{x})^2 = 3178, \ \sum_{i=1}^{4}\sum_{j=1}^{5} (x_{ij} - \bar{x})^2 = 21{,}530.$$

The denominator sum of squares in (9) can now be obtained by means of formula (8). Thus

$$\sum_{i=1}^{4}\sum_{i=1}^{5} (x_{ij} - \bar{x}_{i.} - \bar{x}_{.j} + \bar{x})^2 = 21{,}530 - 5(1286) - 4(3178) = 2388.$$

The value of F in (9) is therefore given by

$$F = \frac{4 \cdot 3 \cdot 3{,}178}{2388} = 16.0$$

$$\nu_1 = 4, \qquad \nu_2 = 12.$$

From Table IX it will be found that the 5 per cent critical value of F corresponding to $\nu_1 = 4$ and $\nu_2 = 12$ is 3.26. Since $F = 16.0$ is in the critical region, the hypothesis that the fertilizers are equally effective is rejected. Fertilizers E and C appear to be superior to the other fertilizers.

The preceding methods can be generalized to include more than two variables of classification. The methods for doing so are quite similar to those just explained. Thus, by means of the analysis of variance technique it is possible to design and analyze very complicated experiments involving several variables of experimental interest.

EXERCISES

Section 2

1. The following data give the yield of a chemical product that resulted from trying 4 different catalysts in the chemical process. Use formula (7) to test to see whether yields are influenced by the catalysts.

I	II	III	IV
36	35	35	34
33	37	39	31
35	36	37	35
34	35	38	32
32	37	39	34
34	36	38	33

2. The following data give the yields of wheat on some experimental plots of ground corresponding to 4 different sulfur treatments for the control of rust. The treatments consisted of dusting before rains, dusting after rains, dusting once each week, and no dusting. Test by means of formula (7) to see if there are significant differences in yields due to the dusting methods.

Dusting Method

Plot	1	2	3	4
1	5.3	4.4	8.4	7.4
2	3.7	5.1	6.0	4.3
3	14.3	5.4	4.9	3.5
4	6.5	12.1	9.5	3.8

3. For the following data on the scores made by pupils taught arithmetic by three different methods, use formula (7) to form the appropriate F test and test for equality of column means.

Methods

116	132	108
117	137	96
138	131	131
100	108	130
125	111	111
130	130	126
134	140	
124		
114		

4. In problem 2 use Student's t test to determine whether dusting had any effect upon yield. That is, combine the data for the first three columns and treat the measurements as though they were a sample of size 12 from a normal population. The last column measurements may be treated as a sample of size 4 from a second normal population.

Section 3

5. The following data represent the number of units of production per day turned out by 5 workmen using 4 different types of machines. (a) Test to see whether the mean

productivity is the same for the 4 different machine types. (b) Test to see whether the 5 men differ with respect to mean productivity. Use formulas (9) and (10) here.

Workman	Machine Type			
	1	2	3	4
1	44	38	47	36
2	46	40	52	43
3	34	36	44	32
4	43	38	46	33
5	38	42	49	39

6. Work problem 2 by the two-variable method to see whether eliminating the variation due to plots will affect the test.

7. For the data of Table 4 in the text, test to see whether there are plot differences in yield.

8. For the data of Table 4, use formula (7) to test whether the fertilizers are equally effective; that is, ignore plot differences.

9. An experiment to determine whether the yield of a chemical process can be increased by changing the temperature of the reaction and by changing the amount of a catalyst was carried out with the following results. The entries in the table represent yields under three temperatures and four catalyst choices. (a) Test to see whether the yield is affected by varying the catalyst. (b) Test to see whether the yield is affected by varying the temperature.

Temper-ature	Catalyst			
	1	2	3	4
A	53	59	58	50
B	57	65	62	60
C	52	62	54	52

CHAPTER 13

Nonparametric Tests

1. INTRODUCTION

With one exception, all the methods that have been presented thus far for testing hypotheses have assumed that the form of the population distribution was known and that a test of some assumption about a parameter of the distribution was to be made. For example, methods were derived for testing the value of μ for a normal distribution and for testing the value of p for a binomial distribution. The one exception occurred in Chapter 11 where the χ^2 distribution was employed to test the compatibility of a set of observed frequencies with a set of expected frequencies. Now, expected frequencies are often obtained without a knowledge of the distribution of the basic underlying variable. For example, in the contingency table problem it was not necessary to know how the two basic variables were distributed in order to test whether those variables were independent.

Situations often arise in which one has no knowledge of the distribution of the variable being studied or in which one knows definitely that it is not of the required type for applying the desired theory. For example, one may know that the variable possesses a distribution far different from a normal distribution, yet one would like to test whether the mean of the distribution has a specified value. Since small sample methods require a normality assumption, the standard method would not be applicable to such a problem when only a small sample is available.

Methods have been designed to take the place of standard tests when the assumptions required by those tests are not satisfied. They are called nonparametric methods because they do not test parameter values of known population types, as in the case of the standard tests in the preceding chapters. Since nonparametric methods are more general than those requiring additional assumptions, it is to be expected that they will not be quite so good as the standard methods when both are applicable. These new methods should therefore be used only when a standard method is not appropriate.

The nonparametric methods that are presented in this chapter were chosen to solve some of the same types of problems solved earlier by parametric methods. There are quite a few other nonparametric tests available for these same problems and for other types of problems as well.

2.   TESTING A MEDIAN

For nonparametric problems related to continuous variables the median is a more natural measure of location for a distribution than the mean. The median has the desirable property that the probability is $\frac{1}{2}$ that a sample value will exceed the population median, regardless of the nature of the distribution. As a result, it is possible to design tests for testing hypothetical values of the median without knowing what the underlying distribution is like. The simplest is the *sign test*. For the purpose of describing this test, consider the following data obtained from testing the breaking strength of ceramic tile manufactured by a new cheaper process: 20, 42, 18, 21, 22, 35, 19, 18, 26, 20, 21, 32, 22, 20, 24. Suppose that experience with the old process produced a median of 25. Then a natural hypothesis to test here is

$$H_0 : \zeta = 25 \quad \text{against} \quad H_1 : \zeta < 25$$

where ζ denotes the median of the distribution.

The first step in applying the sign test is to subtract the postulated median from each observed measurement and then record the sign of the corresponding difference. If 25 is subtracted from each of the foregoing observed values, the following signs will be obtained:

$$- + - - - + - - + - - + - - -$$

The next step is to count the number of + signs, denoted by x, and the total number of signs, denoted by n. Here, $x = 4$ and $n = 15$. If the hypothesis H_0 is true, the probability is $\frac{1}{2}$ that a + sign will be obtained when an observation is taken; consequently, the variable x represents the number of successes in n trials of an experiment for which the probability of success in a single trial is $p = \frac{1}{2}$. The problem has now been reduced to a binomial distribution problem of the type treated in Chapter 8.

If the median of a distribution is smaller than the postulated value, then subtracting the postulated value from a set of observed measurements is likely to produce more negative than positive differences, hence a value of x that is smaller than expected under the hypothesis. Since small values of x favor the alternative hypothesis, the critical region of the test should be in the left tail of the binomial distribution.

Calculations by means of the binomial distribution formula given by (1), Chapter 5, with $n = 15$ and $p = \frac{1}{2}$, yielded the following probabilities:

$$P\{0\} = .000, \qquad P\{3\} = .014,$$
$$P\{1\} = .000, \qquad P\{4\} = .042.$$
$$P\{2\} = .003,$$

When these probabilities are summed, it follows that $P\{x \leq 4\} = .059$. Thus the observed value of $x = 4$ would be in the critical region if a critical region of size $\alpha = .059$ were selected, but it would not be in the critical region if a smaller value of α were selected.

If the normal approximation to the binomial distribution based on the geometrical method shown in section 4, Chapter 5, had been used here, one would have calculated

$$z = \frac{x - np}{\sqrt{npq}} = \frac{4.5 - 7.5}{\sqrt{15 \cdot \frac{1}{2} \cdot \frac{1}{2}}} = -1.55.$$

From Table IV in the appendix, $P\{z < -1.55\} = .06$, which agrees very well with the result obtained by calculating the necessary binomial probabilities.

The problem that was just solved by nonparametric methods corresponds to the first hypothesis testing problem considered in Chapter 8, which consisted of testing whether the mean of a normal distribution had a particular value. The next section considers a nonparametric analogue of the problem of testing whether the means of two normal distributions are equal.

3. TESTING THE DIFFERENCE OF TWO MEDIANS

Since the median is being used as a substitute for the mean as a location parameter in nonparametric problems, it is natural to test the difference of two medians rather than the difference of two means in nonparametric situations. This hypothesis may be written in the form

$$H_0 : \xi_1 = \xi_2,$$

in which ξ_1 and ξ_2 denote the medians of the two populations of interest. The nonparametric test that is introduced to solve this type of problem is called the *rank-sum test*. To illustrate how the test is applied, consider the following data on the number of trials required by rats to learn a certain task for a group of eight treated rats and a group of ten untreated rats:

T	24	28	15	47	23	25	53	20		
U	22	12	30	16	26	14	18	21	16	18

The two samples are first arranged together in order of increasing size. For the foregoing data this yields the following ordering, in which the entries from the treated group have been underlined:

12, 14, 15, 16, 16, 18, 18, 20, 21,
22, 23, 24, 25, 26, 28, 30, 47, 53.

These values are then replaced by their proper ranks to give

1, 2, 3, 4, 5, 6, 7, 8, 9,
10, 11, 12, 13, 14, 15, 16, 17, 18.

The next step is to sum the ranks of the smaller group (here the treated group). If this sum is denoted by R, it follows that the value of R will be given by summing the underlined ranks. For this problem $R = 97$. The final step is to determine whether the value of R lies in the critical region of the test.

The sampling distribution of R, under the assumption that the two population distributions are identical, has been worked out by mathematical methods. The distribution of R depends, of course, upon the sizes of the two samples, which are denoted by n_1 and n_2. Here n_1 denotes the smaller of the two sample sizes. Table VIII in the appendix gives the desired critical values corresponding to various sample sizes for $n_2 \leq 10$. For larger sample sizes, the distribution of R can be approximated satisfactorily by the proper normal distribution. This is the normal distribution with mean and standard deviation given by the formulas

$$\mu_R = \frac{n_1(n_1 + n_2 + 1)}{2},$$

$$\sigma_R = \sqrt{\frac{n_1 n_2(n_1 + n_2 + 1)}{12}}.$$

For the problem being discussed it was expected that the treated rats would require a longer learning period than the untreated rats; therefore, the natural alternative hypothesis here is

$$H_1 : \xi_1 > \xi_2.$$

Under this alternative, R would tend to be larger than under H_0 because R is the sum of the ranks of the ξ_1 group of measurements; consequently, one should choose the critical region under the right tail of the distribution. From Table VIII it will be found that for $n_1 = 8$ and $n_2 = 10$ the probability is .051 that $R \geq 95$. Since $R = 97$ lies in the critical region, the hypothesis is rejected.

If the normal approximation had been used, one would have calculated

$$\mu_R = \frac{8(8 + 10 + 1)}{2} = 76$$

and

$$\sigma_R = \sqrt{\frac{8 \cdot 10(8 + 10 + 1)}{12}} = 11.3.$$

Then one would have calculated

$$z = \frac{R - \mu_R}{\sigma_R} = \frac{97 - 76}{11.3} = 1.86.$$

From Table IV in the appendix it will be found that $P\{z > 1.86\} = .03$; hence H_0 would be rejected. A slightly more accurate approximation could have been obtained by using 96.5 in place of 97 in 2.

When $n_1 = n_2$, one may choose either group as the smaller for which R is to be computed. When the two groups of observations contain one or more common values, ties in ranking will occur. In such situations it suffices to give each set of equal observed values the rank that is the mean of the ranks occupied by them. This modification is not necessary for ties that occur in the same group.

Since the distribution of R was obtained on the assumption that the two population distributions were identical, one is really testing the hypothesis that the two distributions are identical against the alternative that one of them has been shifted to the right. Two populations may have identical medians and yet differ in such a manner as to produce sample values of R that would regularly fall in the critical region of the preceding test. Thus it is not strictly correct to reject the hypothesis $H_0: \xi_1 = \xi_2$ when R falls in the critical region unless one is prepared to assume that the two distributions are identical except possibly for their locations. The foregoing test is often called a slippage test because it determines whether two distributions are identical against the possibility that one of them may have slipped relative to the other.

It is interesting to compare the rank-sum test for this problem with the corresponding parametric test that would be applied here if the two basic variables could be assumed to possess independent normal distributions. Since the sample sizes are rather small this comparison will be made by employing Student's t test as explained in section 3.1 of Chapter 8. This requires the assumption of equal variances for the variables in addition to the normality assumption. Calculations based on the formula of section 3.1 of Chapter 8 yielded the value $t = 2.18$. Since there are $\nu = 16$ degrees of freedom here, Table V shows that this t value is very close to the .025 listed value. This probability is close to the probability value of .03 that was obtained by applying the normal distribution approximation to R for this problem; therefore the two tests differ very little on this problem.

The rank-sum test is known to be excellent for testing slippage. Even when there is justification in assuming that the two variables are independently normally distributed with the same variances, the rank-sum test does nearly as well as the Student t test, which was designed for this type of problem, in the sense of producing small type II errors. This property of being nearly as good as the t test when the latter is justified and being a valid test under all conditions makes the rank-sum test a very attractive test.

The two foregoing nonparametric methods were selected to show how typical parametric problems can be solved by nonparametric methods. There are nonparametric techniques available for most of the parametric problems discussed in the preceding chapters. They are not discussed here because this chapter was intended only to give a glimpse of methods that can be applied when the assumptions needed for the standard method are not satisfied.

EXERCISES

Section 2

1. The following data represent the ages of burglars apprehended during the past year in a certain city. Test the hypothesis that the median age for burglars is 23 years, which is the age obtained from past experience, against the alternative that it is less than 23. In making the test, ignore the 23-year-old individuals. The ages are 24, 16, 18, 19, 18, 17, 26, 28, 23, 22, 19, 21, 21, 23, 17, 15, 21, 36, 18, 38, 19, 22, 23, 26, 21, 33, 21, 17, 26, 18.

2. In an elementary school seventeen pairs of first-grade children were formed on the basis of similarity of intelligence and background. One child of each pair was taught by reading method I and the other by method II. After a period of training, the children were given a reading test with the following results:

Method I	65	68	70	63	64	62	73	75	72	78	64	73	79	80	67	74	82
Method II	63	68	68	60	65	66	72	74	73	70	66	70	80	78	63	74	78

Taking differences of values and ignoring zero values, use the sign test to determine whether the two methods are equally effective.

3. Work problem 2 by means of Student's t test, after taking differences, as explained in Chapter 8, and comment on the two outcomes.

4. Work problem 14, Chapter 8, by means of the sign test and compare your result with that obtained by using the t test.

5. Experience with a manufacturing process has shown that 2 per cent of the parts turned out by certain machines are defective. One of the machines is suspected of being faulty and so a record is kept of its daily percentage of defective parts. This record of percentages is: 2.2, 2.3, 2.1, 1.7, 3.8, 2.5, 1.9, 1.6, 1.4, 2.6, 1.5, 2.8, 2.9, 2.6, 2.5, 3.2, 4.6, 3.3, 3.0, 1.8, 2.6, 2.1, 1.8, 2.4, 2.4, 1.7, 1.6, 2.8, 3.0, 2.4. Use the sign test to test the hypothesis that the median of the distribution is 2.

6. The following data represent the number of red blood cells (million per cubic millimeter) for ten men and ten women. (a) Use the rank-sum test to see whether there is any difference between the sexes with respect to red blood cells. (b) Work the problem by using the normal approximation for this test.

Men	5.02	4.58	5.57	4.52	4.84	5.36	4.27	5.15	4.93	4.72
Women	4.15	4.56	3.89	4.40	4.38	4.20	4.31	4.73	4.26	3.95

7. Explain why the sign test applied to the differences of values in problem 6 would not be a satisfactory test to apply to such data.

8. These data represent the scores made on a grammar test by two groups of students. Group I students graduated from private schools, whereas Group II students are public-school graduates. Use the rank-sum test to test for equality of the medians.

I	25	30	28	34	24	25	13	32	24	30	31	36		
II	44	33	22	8	47	31	40	30	33	35	18	21	35	29	22

9. Work problem 6 by means of Student's t test for the difference of two means, as explained in Chapter 8, and comment on the two outcomes.

10. Work problem 23, Chapter 8, by means of the rank-sum test.

11. The following two sets of values represent the results of sampling 12 males and 15 females and giving them a test that purports to measure one's tolerance level. Use the rank-sum test to test the hypothesis that the two groups possess the same distribution with respect to this measure.

M	25	30	28	34	24	25	13	32	24	30	31	35		
F	44	34	22	8	47	31	40	30	32	35	18	21	35	29	22

Appendix Tables

TABLE I. Squares and Square Roots

N	N²	√N	√10N	N	N²	√N	√10N
1.00	1.0000	1.00000	3.16228	**1.50**	2.2500	1.22474	3.87298
1.01	1.0201	1.00499	3.17805	1.51	2.2801	1.22882	3.88587
1.02	1.0404	1.00995	3.19374	1.52	2.3104	1.23288	3.89872
1.03	1.0609	1.01489	3.20936	1.53	2.3409	1.23693	3.91152
1.04	1.0816	1.01980	3.22490	1.54	2.3716	1.24097	3.92428
1.05	1.1025	1.02470	3.24037	1.55	2.4025	1.24499	3.93700
1.06	1.1236	1.02956	3.25576	1.56	2.4336	1.24900	3.94968
1.07	1.1449	1.03441	3.27109	1.57	2.4649	1.25300	3.96232
1.08	1.1664	1.03923	3.28634	1.58	2.4964	1.25698	3.97492
1.09	1.1881	1.04403	3.30151	1.59	2.5281	1.26095	3.98748
1.10	1.2100	1.04881	3.31662	**1.60**	2.5600	1.26491	4.00000
1.11	1.2321	1.05357	3.33167	1.61	2.5921	1.26886	4.01248
1.12	1.2544	1.05830	3.34664	1.62	2.6244	1.27279	4.02492
1.13	1.2769	1.06301	3.36155	1.63	2.6569	1.27671	4.03733
1.14	1.2996	1.06771	3.37639	1.64	2.6896	1.28062	4.04969
1.15	1.3225	1.07238	3.39116	1.65	2.7225	1.28452	4.06202
1.16	1.3456	1.07703	3.40588	1.66	2.7556	1.28841	4.07431
1.17	1.3689	1.08167	3.42053	1.67	2.7889	1.29228	4.08656
1.18	1.3924	1.08628	3.43511	1.68	2.8224	1.29615	4.09878
1.19	1.4161	1.09087	3.44964	1.69	2.8561	1.30000	4.11096
1.20	1.4400	1.09545	3.46410	**1.70**	2.8900	1.30384	4.12311
1.21	1.4641	1.10000	3.47851	1.71	2.9241	1.30767	4.13521
1.22	1.4884	1.10454	3.49285	1.72	2.9584	1.31149	4.14729
1.23	1.5129	1.10905	3.50714	1.73	2.9929	1.31529	4.15933
1.24	1.5376	1.11355	3.52136	1.74	3.0276	1.31909	4.17133
1.25	1.5625	1.11803	3.53553	1.75	3.0625	1.32288	4.18330
1.26	1.5876	1.12250	3.54965	1.76	3.0976	1.32665	4.19524
1.27	1.6129	1.12694	3.56371	1.77	3.1329	1.33041	4.20714
1.28	1.6384	1.13137	3.57771	1.78	3.1684	1.33417	4.21900
1.29	1.6641	1.13578	3.59166	1.79	3.2041	1.33791	4.23084
1.30	1.6900	1.14018	3.60555	**1.80**	3.2400	1.34164	4.24264
1.31	1.7161	1.14455	3.61939	1.81	3.2761	1.34536	4.25441
1.32	1.7424	1.14891	3.63318	1.82	3.3124	1.34907	4.26615
1.33	1.7689	1.15326	3.64692	1.83	3.3489	1.35277	4.27785
1.34	1.7956	1.15758	3.66060	1.84	3.3856	1.35647	4.28952
1.35	1.8225	1.16190	3.67423	1.85	3.4225	1.36015	4.30116
1.36	1.8496	1.16619	3.68782	1.86	3.4596	1.36382	4.31277
1.37	1.8769	1.17047	3.70135	1.87	3.4969	1.36748	4.32435
1.38	1.9044	1.17473	3.71484	1.88	3.5344	1.37113	4.33590
1.39	1.9321	1.17898	3.72827	1.89	3.5721	1.37477	4.34741
1.40	1.9600	1.18322	3.74166	**1.90**	3.6100	1.37840	4.35890
1.41	1.9881	1.18743	3.75500	1.91	3.6481	1.38203	4.37035
1.42	2.0164	1.19164	3.76829	1.92	3.6864	1.38564	4.38178
1.43	2.0449	1.19583	3.78153	1.93	3.7249	1.38924	4.39318
1.44	2.0736	1.20000	3.79473	1.94	3.7636	1.39284	4.40454
1.45	2.1025	1.20416	3.80789	1.95	3.8025	1.39642	4.41588
1.46	2.1316	1.20830	3.82099	1.96	3.8416	1.40000	4.42719
1.47	2.1609	1.21244	3.83406	1.97	3.8809	1.40357	4.43847
1.48	2.1904	1.21655	3.84708	1.98	3.9204	1.40712	4.44972
1.49	2.2201	1.22066	3.86005	1.99	3.9601	1.41067	4.46094
1.50	2.2500	1.22474	3.87298	**2.00**	4.0000	1.41421	4.47214
N	N²	√N	√10N	N	N²	√N	√10N

TABLE I. (Continued)

N	N²	√N̄	√10N̄	N	N²	√N̄	√10N̄
2.00	4.0000	1.41421	4.47214	**2.50**	6.2500	1.58114	5.00000
2.01	4.0401	1.41774	4.48330	2.51	6.3001	1.58430	5.00999
2.02	4.0804	1.42127	4.49444	2.52	6.3504	1.58745	5.01996
2.03	4.1209	1.42478	4.50555	2.53	6.4009	1.59060	5.02991
2.04	4.1616	1.42829	4.51664	2.54	6.4516	1.59374	5.03984
2.05	4.2025	1.43178	4.52769	2.55	6.5025	1.59687	5.04975
2.06	4.2436	1.43527	4.53872	2.56	6.5536	1.60000	5.05964
2.07	4.2849	1.43875	4.54973	2.57	6.6049	1.60312	5.06952
2.08	4.3264	1.44222	4.56070	2.58	6.6564	1.60624	5.07937
2.09	4.3681	1.44568	4.57165	2.59	6.7081	1.60935	5.08920
2.10	4.4100	1.44914	4.58258	**2.60**	6.7600	1.61245	5.09902
2.11	4.4521	1.45258	4.59347	2.61	6.8121	1.61555	5.10882
2.12	4.4944	1.45602	4.60435	2.62	6.8644	1.61864	5.11859
2.13	4.5369	1.45945	4.61519	2.63	6.9169	1.62173	5.12835
2.14	4.5796	1.46287	4.62601	2.64	6.9696	1.62481	5.13809
2.15	4.6225	1.46629	4.63681	2.65	7.0225	1.62788	5.14782
2.16	4.6656	1.46969	4.64758	2.66	7.0756	1.63095	5.15752
2.17	4.7089	1.47309	4.65833	2.67	7.1289	1.63401	5.16720
2.18	4.7524	1.47648	4.66905	2.68	7.1824	1.63707	5.17687
2.19	4.7961	1.47986	4.67974	2.69	7.2361	1.64012	5.18652
2.20	4.8400	1.48324	4.69042	**2.70**	7.2900	1.64317	5.19615
2.21	4.8841	1.48661	4.70106	2.71	7.3441	1.64621	5.20577
2.22	5.9284	1.48997	4.71169	2.72	7.3984	1.64924	5.21536
2.23	4.9729	1.49332	4.72229	2.73	7.4529	1.65227	5.22494
2.24	5.0176	1.49666	4.73286	2.74	7.5076	1.65529	5.23450
2.25	5.0625	1.50000	4.74342	2.75	7.5625	1.65831	5.24404
2.26	5.1076	1.50333	4.75395	2.76	7.6176	1.66132	5.25357
2.27	5.1529	1.50665	4.76445	2.77	7.6729	1.66433	5.26308
2.28	5.1984	1.50997	4.77493	2.78	7.7284	1.66733	5.27257
2.29	5.2441	1.51327	4.78539	2.79	7.7841	1.67033	5.28205
2.30	5.2900	1.51658	4.79583	**2.80**	7.8400	1.67332	5.29150
2.31	5.3361	1.51987	4.80625	2.81	7.8961	1.67631	5.30094
2.32	5.3824	1.52315	4.81664	2.82	7.9524	1.67929	5.31037
2.33	5.4289	1.52643	4.82701	2.83	8.0089	1.68226	5.31977
2.34	5.4756	1.52971	4.83735	2.84	8.0656	1.68523	5.32917
2.35	5.5225	1.53297	4.84768	2.85	8.1225	1.68819	5.33854
2.36	5.5696	1.53623	4.85798	2.86	8.1796	1.69115	5.34790
2.37	5.6169	1.53948	4.86826	2.87	8.2369	1.69411	5.35724
2.38	5.6644	1.54272	4.87852	2.88	8.2944	1.69706	5.36656
2.39	5.7121	1.54596	4.88876	2.89	8.3521	1.70000	5.37587
2.40	5.7600	1.54919	4.89898	**2.90**	8.4100	1.70294	5.38516
2.41	5.8081	1.55252	4.90918	2.91	8.4681	1.70587	5.39444
2.42	5.8564	1.55563	4.91935	2.92	8.5264	1.70880	5.40370
2.43	5.9049	1.55885	4.92950	2.93	8.5849	1.71172	5.41295
2.44	5.9536	1.56205	4.93964	2.94	8.6436	1.71464	5.42218
2.45	6.0025	1.56525	4.94975	2.95	8.7025	1.71756	5.43139
2.46	6.0516	1.56844	4.95984	2.96	8.7616	1.72047	5.44059
2.47	6.1009	1.57162	4.96991	2.97	8.8209	1.72337	5.44977
2.48	6.1054	1.57480	4.97996	2.98	8.8804	1.72627	5.45894
2.49	6.2001	1.57797	4.98999	2.99	8.9401	1.72916	5.46809
2.50	6.2500	1.58114	5.00000	**3.00**	9.0000	1.73205	5.47723
N	N²	√N̄	√10N̄	N	N²	√N̄	√10N̄

TABLE I. (*Continued*)

N	N²	√N	√10N	N	N²	√N	√10N
3.00	9.0000	1.73205	5.47723	**3.50**	12.2500	1.87083	5.91608
3.01	9.0601	1.73494	5.48635	3.51	12.3201	1.87350	5.92453
3.02	9.1204	1.73781	5.49545	3.52	12.3904	1.87617	5.93296
3.03	9.1809	1.74069	5.50454	3.53	12.4609	1.87883	5.94138
3.04	9.2416	1.74356	5.51362	3.54	12.5316	1.88149	5.94979
3.05	9.3025	1.74642	5.52268	3.55	12.6025	1.88414	5.95819
3.06	9.3636	1.74929	5.53173	3.56	12.6736	1.88680	5.96657
3.07	9.4249	1.75214	5.54076	3.57	12.7449	1.88944	5.97495
3.08	9.4864	1.75499	5.54977	3.58	12.8164	1.89209	5.98331
3.09	9.5481	1.75784	5.55878	3.59	12.8881	1.89473	5.99166
3.10	9.6100	1.76068	5.56776	**3.60**	12.9600	1.89737	6.00000
3.11	9.6721	1.76352	5.57674	3.61	13.0321	1.90000	6.00833
3.12	9.7344	1.76635	5.58570	3.62	13.1044	1.90263	6.01664
3.13	9.7969	1.76918	5.59464	3.63	13.1769	1.90526	6.02495
3.14	9.8596	1.77200	5.60357	3.64	13.2496	1.90788	6.03324
3.15	9.9225	1.77482	5.61249	3.65	13.3225	1.91050	6.04152
3.16	9.9856	1.77764	5.62139	3.66	13.3956	1.91311	6.04949
3.17	10.0489	1.78045	5.63028	3.67	13.4689	1.91572	6.05805
3.18	10.1124	1.78326	5.63915	3.68	13.5424	1.91833	6.06630
3.19	10.1761	1.78606	5.64801	3.69	13.6161	1.92094	6.07454
3.20	10.2400	1.78885	5.65685	**3.70**	13.6900	1.92354	6.08276
3.21	10.3041	1.79165	5.66569	3.71	13.7641	1.92614	6.09098
3.22	10.3684	1.79444	5.67450	3.72	13.8384	1.92873	6.09918
3.23	10.4329	1.79722	5.68331	3.73	13.9129	1.93132	6.10737
3.24	10.4976	1.80000	5.69210	3.74	13.9876	1.93391	6.11555
3.25	10.5625	1.80278	5.70088	3.75	14.0625	1.93649	6.12372
3.26	10.6276	1.80555	5.70964	3.76	14.1376	1.93907	6.13188
3.27	10.6929	1.80831	5.71839	3.77	14.2129	1.94165	6.14003
3.28	10.7584	1.81108	5.72713	3.78	14.2884	1.94422	6.14817
3.29	10.8241	1.81384	5.73585	3.79	14.3641	1.94679	6.15630
3.30	10.8900	1.81659	5.74456	**3.80**	14.4400	1.94936	6.16441
3.31	10.9561	1.81934	5.75326	3.81	14.5161	1.95192	6.17252
3.32	10.0224	1.82209	5.76194	3.82	14.5924	1.95448	6.18061
3.33	11.0889	1.82483	5.77062	3.83	14.6689	1.95704	6.18870
3.34	11.1556	1.82757	5.77927	3.84	14.7456	1.95959	6.19677
3.35	11.2225	1.83030	5.78792	3.85	14.8225	1.96214	6.20484
3.36	11.2896	1.83303	5.79655	3.86	14.8996	1.96469	6.21289
3.37	11.3569	1.83576	5.80517	3.87	14.9769	1.96723	6.22093
3.38	11.4244	1.83848	5.81378	3.88	15.0544	1.96977	6.22896
3.39	11.4921	1.84120	5.82237	3.89	15.1321	1.97231	6.23699
3.40	11.5600	1.84391	5.83095	**3.90**	51.2100	1.97484	6.24500
3.41	11.6281	1.84662	5.83952	3.91	15.2881	1.97737	6.25300
3.42	11.6964	1.84932	5.84808	3.92	15,3664	1.97990	6.26099
3.43	11.7649	1.85203	5.85662	3.93	15.4449	1.98242	6.26897
3.44	11.8336	1.85472	5.86515	3.94	15.5236	1.98494	6.27694
3.45	11.9025	1.85742	5.87367	3.95	15.6025	1.98746	6.28490
3.46	11.9716	1.86011	5.88218	3.96	15.6816	1.98997	6.29285
3.47	12.0409	1.86279	5.89067	3.97	15.7609	1.99249	6.30079
3.48	12.1104	1.86548	5.89915	3.98	15.8404	1.99499	6.30872
3.49	12.1801	1.86815	5.90762	3.99	15.9201	1.99750	6.31644
3.50	12.2500	1.87083	5.91608	**4.00**	16.0000	2.00000	6.32456
N	N²	√N	√10N	N	N²	√N	√10N

TABLE I. (*Continued*)

N	N^2	$\sqrt{N}$	$\sqrt{10N}$	N	N^2	$\sqrt{N}$	$\sqrt{10N}$
4.00	16.0000	2.00000	6.32456	**4.50**	20.2500	2.12132	6.70820
4.01	16.0801	2.00250	6.33246	4.51	20.3401	2.12368	6.71565
4.02	16.1604	2.00499	6.34035	4.52	20.4304	2.12603	6.72309
4.03	16.2409	2.00749	6.34823	4.53	20.5209	2.12838	6.73053
4.04	16.3216	2.00998	6.35610	4.54	20.6116	2.13073	6.73795
4.05	16.4025	2.01246	6.36396	4.55	20.7025	2.13307	6.74537
4.06	16.4836	2.01494	6.37181	4.56	20.7936	2.13542	6.75278
4.07	16.5649	2.01742	6.37966	4.57	20.8849	2.13776	6.76018
4.08	16.6464	2.01990	6.38749	4.58	20.9764	2.14009	6.76757
4.09	16.7281	2.02237	6.39531	4.59	21.0681	2.14243	6.77495
4.10	16.8100	2.02485	6.40312	**4.60**	21.1600	2.14476	6.78233
4.11	16.8921	2.02731	6.41093	4.61	21.2521	2.14709	6.78970
4.12	16.9744	2.02978	6.41872	4.62	21.3444	2.14942	6.79706
4.13	17.0569	2.03224	6.42651	4.63	21.4369	2.15174	6.80441
4.14	17.1396	2.03470	6.43428	4.64	21.5296	2.15407	6.81175
4.15	17.2225	2.03715	6.44205	4.65	21.6225	2.15639	6.81909
4.16	17.3056	2.03961	6.44981	4.66	21.7156	2.15870	6.82642
4.17	17.3889	2.04206	6.45755	4.67	21.8089	2.16102	6.83374
4.18	17.4724	2.04450	6.46529	4.68	21.9024	2.16333	6.84105
4.19	17.5561	2.04695	6.47302	4.69	21.9961	2.16564	6.84836
4.20	17.6400	2.04939	6.48074	**4.70**	22.0900	2.16795	6.85565
4.21	17.7241	2.05183	6.48845	4.71	22.1841	2.17025	6.86294
4.22	17.8084	2.05426	6.49615	4.72	22.2784	2.17256	6.87023
4.23	17.8929	2.05670	6.50384	4.73	22.3729	2.17486	6.87750
4.24	17.9776	2.05913	6.51153	4.74	22.4676	2.17715	6.88477
4.25	18.0625	2.06155	6.51920	4.75	22.5625	2.17945	6.89202
4.26	18.1476	2.06398	6.52687	4.76	22.6576	2.18174	6.89928
4.27	18.2329	2.06640	6.53452	4.77	22.7529	2.18403	6.90652
4.28	18.3184	2.06882	6.54217	4.78	22.8484	2.18632	6.91375
4.29	18.4041	2.07123	6.54981	4.79	22.9441	2.18861	6.92098
4.30	18.4900	2.07364	6.55744	**4.80**	23.0400	2.19089	6.92820
4.31	18.5761	2.07605	6.66506	4.81	23.1361	2.19317	6.93542
4.32	18.6624	2.07846	6.57267	4.82	23.2324	2.19545	6.94262
4.33	18.7489	2.08087	6.58027	4.83	23.3289	2.19773	6.94982
4.34	18.8356	2.08327	6.58787	4.84	23.4256	2.20000	6.95701
4.35	18.9225	2.08567	6.59545	4.85	23.5225	2.20227	6.96419
4.36	19.0096	2.08806	6.60303	4.86	23.6196	2.20454	6.97137
4.37	19.0969	2.09045	6.61060	4.87	23.7169	2.20681	6.97854
4.38	19.1844	2.09284	6.61816	4.88	23.8144	2.20907	6.98570
4.39	19.2721	2.09523	6.62571	4.89	23.9121	2.21133	6.99285
4.40	19.3600	2.09762	6.63325	**4.90**	24.0100	2.21359	7.00000
4.41	19.4481	2.10000	6.64078	4.91	24.1081	2.21585	7.00714
4.42	19.5364	2.10238	6.64831	4.92	24.2064	2.21811	7.01427
4.43	19.6249	2.10476	6.65582	4.93	24.3049	2.22036	7.02140
4.44	19.7136	2.10713	6.66333	4.94	24.4036	2.22261	7.02851
4.45	19.8025	2.10950	6.67083	4.95	24.5025	2.22486	7.03562
4.46	19.8916	2.11187	6.67832	4.96	24.6016	2.22711	7.04273
4.47	19.9809	2.11424	6.68581	4.97	24.7009	2.22935	7.04982
4.48	20.0704	2.11660	6.69328	4.98	24.8004	2.23159	7.05691
4.49	20.1601	2.11896	6.70075	4.99	24.9001	2.23383	7.06399
4.50	20.2500	2.12132	6.70820	**5.00**	25.0000	2.23607	7.07107
N	N^2	$\sqrt{N}$	$\sqrt{10N}$	N	N^2	$\sqrt{N}$	$\sqrt{10N}$

TABLE I. *(Continued)*

N	N²	$\sqrt{N}$	$\sqrt{10N}$	N	N²	$\sqrt{N}$	$\sqrt{10N}$
5.00	25.0000	2.23607	7.07107	**5.50**	30.2500	2.34521	7.41620
5.01	25.1001	2.23830	7.07814	5.51	30.3601	2.34734	7.42294
5.02	25.2004	2.24054	7.08520	5.52	30.4704	2.34947	7.42967
5.03	25.3009	2.24277	7.09225	5.53	30.5809	2.35160	7.43640
5.04	25.4016	2.24499	7.09930	5.54	30.6916	2.35372	7.44312
5.05	25.5025	2.24722	7.10634	5.55	30.8025	2.35584	7.44983
5.06	25.6036	2.24944	7.11337	5.56	30.9136	2.35797	7.45654
5.07	25.7049	2.25167	7.12039	5.57	31.0249	2.36008	7.46324
5.08	25.8064	2.25389	7.12741	5.58	31.1364	2.36220	7.46994
5.09	25.9081	2.25610	7.13442	5.59	31.2481	2.36432	7.47663
5.10	26.0100	2.25832	7.14143	**5.60**	31.3600	2.36643	7.48331
5.11	26.1121	2.26053	7.14843	5.61	31.4721	2.36854	7.48999
5.12	26.2144	2.26274	7.15542	5.62	31.5844	2.37065	7.49667
5.13	26.3169	2.26495	7.16240	5.63	31.6969	2.37276	7.50333
5.14	26.4196	2.26716	7.16938	5.64	31.8096	2.37487	7.50999
5.15	26.5225	2.26936	7.17635	5.65	31.9225	2.37697	7.51665
5.16	26.6256	2.27156	7.18331	5.66	32.0356	2.37908	7.52330
5.17	26.7289	2.27376	7.19027	5.67	32.1489	2.38118	7.52994
5.18	26.8324	2.27596	7.19722	5.68	32.2624	2.38328	7.53658
5.19	26.9361	2.27816	7.20417	5.69	32.3761	2.38537	7.54321
5.20	27.0400	2.28035	7.21110	**5.70**	32.4900	2.38747	7.54983
5.21	27.1441	2.28254	7.21803	5.71	32.6041	2.38956	7.55645
5.22	27.2484	2.28473	7.22496	5.72	32.7184	2.39165	7.56307
5.23	27.3529	2.28692	7.23187	5.73	32.8329	2.39374	7.56968
5.24	27.4576	2.28910	7.23838	5.74	32.9476	2.39583	7.57628
5.25	27.5625	2.29129	7.24569	5.75	33.0625	2.39792	7.58288
5.26	27.6676	2.29347	7.25259	5.76	33.1776	2.40000	7.58947
5.27	27.7729	2.29565	7.25948	5.77	33.2929	2.40208	7.59605
5.28	27.8784	2.29783	7.26636	5.78	33.4084	2.40416	7.60263
5.29	27.9841	2.30000	7.27324	5.79	33.5241	2.40624	7.60920
5.30	28.0900	2.30217	7.28011	**5.80**	33.6400	2.40832	7.61577
5.31	28.1961	2.30434	7.28697	5.81	33.7561	2.41039	7.62234
5.32	28.3024	2.30651	7.29383	5.82	33.8724	2.41247	7.62889
5.33	28.4089	2.30868	7.30068	5.83	33.9889	2.41454	7.63544
5.34	28.5156	2.31084	7.30753	5.84	34.1056	2.41661	7.64199
5.35	28.6225	2.31301	7.31437	5.85	34.2225	2.41868	7.64853
5.36	28.7296	2.31517	7.32120	5.86	34.3396	2.42074	7.65506
5.37	28.8369	2.31733	7.32803	5.87	34.4569	2.42281	7.66159
5.38	28.9444	2.31948	7.33485	5.88	34.5744	2.42487	7.66812
5.39	29.0521	2.32164	7.34166	5.89	34.6921	2.42693	7.67463
5.40	29.1600	2.32379	7.34847	**5.90**	34.8100	2.42899	7.68115
5.41	29.2681	2.32594	7.35527	5.91	34.9281	2.43105	7.68765
5.42	29.3764	2.32809	7.36206	5.92	35.0464	2.43311	7.69415
5.43	29.4849	2.33024	7.36885	5.93	35.1649	2.43516	7.70065
5.44	29.5936	2.33238	7.37564	5.94	35.2836	2.43721	7.70714
5.45	29.7025	2.33452	7.38241	5.95	35.4025	2.43926	7.71362
5.46	29.8116	2.33666	7.38918	5.96	35.5216	2.44131	7.72010
5.47	29.9209	2.33880	7.39594	5.97	35.6409	2.44336	7.72658
5.48	30.0304	2.34094	7.40270	5.98	35.7604	2.44540	7.73305
5.49	30.1401	2.34307	7.40945	5.99	35.8801	2.44745	7.73951
5.50	30.2500	2.34521	7.41620	**6.00**	36.0000	2.44949	7.74597
N	N²	$\sqrt{N}$	$\sqrt{10N}$	N	N²	$\sqrt{N}$	$\sqrt{10N}$

TABLE I. (*Continued*)

N	N²	√N	√10N		N	N²	√N	√10N
6.00	36.0000	2.44949	7.74597		**6.50**	42.2500	2.54951	8.06226
6.01	36.1201	2.45153	7.75242		6.51	42.3801	2.55147	8.06846
6.02	36.2404	2.45357	7.75887		6.52	42.5104	2.55343	8.07465
6.03	36.3609	2.45561	7.76531		6.53	42.6409	2.55539	8.08084
6.04	36.4816	2.45764	7.77174		6.54	42.7716	2.55734	8.08703
6.05	36.6025	2.45967	7.77817		6.55	42.9025	2.55930	8.09321
6.06	36.7236	2.46171	7.78460		6.56	43.0336	2.56125	8.09938
6.07	36.8449	2.46374	7.79102		6.57	43.1649	2.56320	8.10555
6.08	36.9664	2.46577	7.79744		6.58	43.2964	2.56515	8.11172
6.09	37.0881	2.46779	7.80385		6.59	43.4281	2.56710	8.11788
6.10	37.2100	2.46982	7.81025		**6.60**	43.5600	2.56905	8.12404
6.11	37.3321	2.47184	7.81665		6.61	43.6921	2.57099	8.13019
6.12	37.4544	2.47386	7.82304		6.62	43.8244	2.57294	8.13634
6.13	37.5769	2.47588	7.82943		6.63	43.9569	2.57488	8.14248
6.14	37.6996	2.47790	7.83582		6.64	44.0896	2.57682	8.14862
6.15	37.8225	2.47992	7.84219		6.65	44.2225	2.57876	8.15475
6.16	37.9456	2.48193	7.84857		6.66	44.3556	2.58070	8.16088
6.17	38.0689	2.48395	7.85493		6.67	44.4889	2.58263	8.16701
6.18	38.1924	2.48596	7.86130		6.68	44.6224	2.58457	8.17313
6.19	38.3161	2.48797	7.86766		6.69	44.7561	2.58650	8.17924
6.20	38.4400	2.48998	7.87401		**6.70**	44.8900	2.58844	8.18535
6.21	38.5641	2.49199	7.88036		6.71	45.0241	2.59037	8.19146
6.22	38.6884	2.49399	7.88670		6.72	45.1584	2.59230	8.19756
6.23	38.8129	2.49600	7.89303		6.73	45.2929	2.59422	8.20366
6.24	38.9376	2.49800	7.89937		6.74	45.4276	2.59615	8.20975
6.25	39.0625	2.50000	7.90569		6.75	45.5625	2.59808	8.21584
6.26	39.1876	2.50200	7.91202		6.76	45.6976	2.60000	8.22192
6.27	39.3129	2.50400	7.91833		6.77	45.8329	2.60192	8.22800
6.28	39.4384	2.50599	7.92465		6.78	45.9684	2.60384	8.23408
6.29	39.5641	2.50799	7.93095		6.79	46.1041	2.60576	8.24015
6.30	39.6900	2.50998	7.93725		**6.80**	46.2400	2.60768	8.24621
6.31	39.8161	2.51197	7.94355		6.81	46.3761	2.60960	8.25227
6.32	39.9424	2.51396	7.94984		6.82	46.5124	2.61151	8.25833
6.33	40.0689	2.51595	7.95613		6.83	46.6489	2.61343	8.26438
6.34	40.1956	2.51794	7.96241		6.84	46.7856	2.61534	8.27043
6.35	40.3225	2.51992	7.96869		6.85	46.9225	2.61725	8.27647
6.36	40.4496	2.52190	7.97496		6.86	47.0596	2.61916	8.28251
6.37	40.5769	2.52389	7.98123		6.87	47.1969	2.62107	8.28855
6.38	40.7044	2.52587	7.98749		6.88	47.3344	2.62298	8.29458
6.39	40.8321	2.52784	7.99375		6.89	47.4721	2.62488	8.30060
6.40	40.9600	2.52982	8.00000		**6.90**	47.6100	2.62679	8.30662
6.41	41.0881	2.53180	8.00625		6.91	47.7481	2.62869	8.31264
6.42	41.2164	2.53377	8.01249		6.92	47.8864	2.63059	8.31865
6.43	41.3449	2.53574	8.01873		6.93	48.0249	2.63249	8.32466
6.44	41.4736	2.53772	8.02496		6.94	48.1636	2.63439	8.33067
6.45	41.6025	2.53969	8.03119		6.95	48.3025	2.63629	8.33667
6.46	41.7316	2.54165	8.03741		6.96	48.4416	2.63818	8.34266
6.47	41.8609	2.54362	8.04363		6.97	48.5809	2.64008	8.34865
6.48	41.9904	2.54558	8.04984		6.98	48.7204	2.64197	8.35464
6.49	42.1201	2.54755	8.05605		6.99	48.8601	2.64386	8.36062
6.50	42.2500	2.54951	8.06226		**7.00**	49.0000	2.64575	8.36660
N	N²	√N	√10N		N	N²	√N	√10N

TABLE I. (Continued)

N	N²	√N	√10N	N	N²	√N	√10N
7.00	49.0000	2.64575	8.36660	**7.50**	56.2500	2.73861	8.66025
7.01	49.1401	2.64764	8.37257	7.51	56.4001	2.74044	8.66603
7.02	49.2804	2.64953	8.37854	7.52	56.5504	2.74226	8.67179
7.03	49.4209	2.65141	8.38451	7.53	56.7009	2.74408	8.67756
7.04	49.5616	2.65330	8.39047	7.54	56.8516	2.74591	8.68332
7.05	49.7025	2.65518	8.39643	7.55	57.0025	2.74773	8.68907
7.06	49.8436	2.65707	8.40238	7.56	57.1536	2.74955	8.69483
7.07	49.9849	2.65895	8.40833	7.57	57.3049	2.75136	8.70057
7.08	50.1264	2.66083	8.41427	7.58	57.4564	2.75318	8.70632
7.09	50.2681	2.66271	8.42021	7.59	57.6081	2.75500	8.71206
7.10	50.4100	2.66458	8.42615	**7.60**	57.7600	2.75681	8.71780
7.11	50.5521	2.66646	8.43208	7.61	57.9121	2.75862	8.72353
7.12	50.6944	2.66833	8.43801	7.62	58.0644	2.76043	8.72926
7.13	50.8369	2.67021	8.44393	7.63	58.2169	2.76225	8.73499
7.14	50.9796	2.67208	8.44985	7.64	58.3696	2.76405	8.74071
7.15	51.1225	2.67395	8.45577	7.65	58.5225	2.76586	8.74643
7.16	51.2656	2.67582	8.46168	7.66	58.6756	2.76767	8.75214
7.17	51.4089	2.67769	8.46759	7.67	58.8289	2.76948	8.75785
7.18	51.5524	2.67955	8.47349	7.68	58.9824	2.77128	8.76356
7.19	51.6961	2.68142	8.47939	7.69	59.1361	2.77308	8.76926
7.20	51.8400	2.68328	8.48528	**7.70**	59.2900	2.77489	8.77496
7.21	51.9841	2.68514	8.49117	7.71	59.4441	2.77669	8.78066
7.22	52.1284	2.68701	8.49706	7.72	59.5984	2.77849	8.78635
7.23	52.2729	2.68887	8.50294	7.73	59.7529	2.78029	8.79204
7.24	52.4176	2.69072	8.50882	7.74	59.9076	2.78209	8.79773
7.25	52.5625	2.69258	8.51469	7.75	60.0625	2.78388	8.80341
7.26	52.7076	2.69444	8.52056	7.76	60.2176	2.78568	8.80909
7.27	52.8529	2.69629	8.52643	7.77	60.3729	2.78747	8.81476
7.28	52.9984	2.69815	8.53229	7.78	60.5284	2.78927	8.82043
7.29	53.1441	2.70000	8.53815	7.79	60.6841	2.79106	8.82610
7.30	53.2900	2.70185	8.54400	**7.80**	60.8400	2.79285	8.83176
7.31	53.4361	2.70370	8.54985	7.81	60.9961	2.79464	8.83742
7.32	53.5824	2.70555	8.55570	7.82	61.1524	2.79643	8.84308
7.33	53.7289	2.70740	8.56154	7.83	61.3089	2.79821	8.84873
7.34	53.8756	2.70924	8.56738	7.84	61.4656	2.80000	8.85438
7.35	54.0225	2.71109	8.57321	7.85	61.6225	2.80179	8.86002
7.36	54.1696	2.71293	8.57904	7.86	61.7796	2.80357	8.86566
7.37	54.3169	2.71477	8.58487	7.87	61.9369	2.80535	8.87130
7.38	54.4644	2.71662	8.59069	7.88	62.0944	2.80713	8.87694
7.39	54.6121	2.71846	8.59651	7.89	62.2521	2.80891	8.88257
7.40	54.7600	2.72029	8.60233	**7.90**	62.4100	2.81069	8.88819
7.41	54.9081	2.72213	8.60814	7.91	62.5681	2.81247	8.89382
7.42	55.0564	2.72397	8.61394	7.92	62.7264	2.81425	8.89944
7.43	55.2049	2.72580	8.61974	7.93	62.8849	2.81603	8.90505
7.44	55.3536	2.72764	8.62554	7.94	63.0436	2.81780	8.91067
7.45	55.5025	2.72947	8.63134	7.95	63.2025	2.81957	8.91628
7.46	55.6516	2.73130	8.63713	7.96	63.3616	2.82135	8.92188
7.47	55.8009	2.73313	8.64292	7.97	63.5209	2.82312	8.92749
7.48	55.9504	2.73496	8.64870	7.98	63.6804	2.82489	8.93308
7.49	56.1001	2.73679	8.65448	7.99	63.8401	2.82666	8.93868
7.50	56.2500	2.73861	8.66025	**8.00**	64.0000	2.82843	8.94427
N	N²	√N	√10N	N	N²	√N	√10N

TABLE I. (Continued)

N	N²	√N	√10N	N	N²	√N	√10N
8.00	64.0000	2.82843	8.94427	**8.50**	72.2500	2.91548	9.21954
8.01	64.1601	2.83019	8.94986	8.51	72.4201	2.91719	9.22497
8.02	64.3204	2.83196	8.95545	8.52	72.5904	2.91890	9.23038
8.03	64.4809	2.83373	8.96103	8.53	72.7609	2.92062	9.23580
8.04	64.6416	2.83549	8.96660	8.54	72.9316	2.92233	9.24121
8.05	64.8025	2.83725	8.97218	8.55	73.1025	2.92404	9.24662
8.06	64.9636	2.83901	8.97775	8.56	73.2736	2.92575	9.25203
8.07	65.1249	2.84077	8.98332	8.57	73.4449	2.92746	9.25743
8.08	65.2864	2.84253	8.98888	8.58	73.6164	2.92916	9.26283
8.09	65.4481	2.84429	8.99444	8.59	73.7881	2.93087	9.26823
8.10	65.6100	2.84605	9.00000	**8.60**	73.9600	2.93258	9.27362
8.11	65.7721	2.84781	9.00555	8.61	74.1321	2.93428	9.27901
8.12	65.9344	2.84956	9.01110	8.62	74.3044	2.93598	9.28440
8.13	66.0969	2.85132	9.01665	8.63	74.4769	2.93769	9.28978
8.14	66.2596	2.85307	9.02219	8.64	74.6496	2.93939	9.29516
8.15	66.4225	2.85482	9.02774	8.65	74.8225	2.94109	9.30054
8.16	66.5856	2.85657	9.03327	8.66	74.9956	2.94279	9.30591
8.17	66.7489	2.85832	9.03881	8.67	75.1689	2.94449	9.31128
8.18	66.9124	2.86007	9.04434	8.68	75.3424	2.94618	9.31665
8.19	67.0761	2.86182	9.04986	8.69	75.5161	2.94788	9.32202
8.20	67.2400	2.86356	9.05539	**8.70**	75.6900	2.94958	9.32738
8.21	67.4041	2.86531	9.06091	8.71	75.8641	2.95127	9.33274
8.22	67.5684	2.86705	9.06642	8.72	76.0384	2.95296	9.33809
8.23	67.7329	2.86880	9.07193	8.73	76.2129	2.95466	9.34345
8.24	67.8976	2.87054	9.07744	8.74	76.3876	2.95635	9.34880
8.25	68.0625	2.87228	9.08295	8.75	76.5625	2.95804	9.35414
8.26	68.2276	2.87402	9.08845	8.76	76.7376	2.95973	9.35949
8.27	68.3929	2.87576	9.09395	8.77	76.9129	2.96142	9.36483
8.28	68.5584	2.87750	9.09945	8.78	77.0884	2.96311	9.37017
8.29	68.7241	2.87924	9.10494	8.79	77.2641	2.96479	9.37550
8.30	68.8900	2.88097	9.11045	**8.80**	77.4400	2.96648	9.38083
8.31	69.0561	2.88271	9.11592	8.81	77.6161	2.96816	9.38616
8.32	69.2224	2.88444	9.12140	8.82	77.7924	2.96985	9.39149
8.33	69.3889	2.88617	9.12688	8.83	77.9689	2.97153	9.39681
8.34	69.5556	2.88791	9.13236	8.84	78.1456	2.97321	9.40213
8.35	69.7225	2.88964	9.13783	8.85	78.3225	2.97489	9.40744
8.36	69.8896	2.89137	9.14330	8.86	78.4996	2.97658	9.41276
8.37	70.0569	2.89310	9.14877	8.87	78.6769	2.97825	9.41807
8.38	70.2244	2.89482	9.15423	8.88	78.8544	2.97993	9.42338
8.39	70.3921	2.89655	9.15969	8.89	79.0321	2.98161	9.42868
8.40	70.5600	2.89828	9.16515	**8.90**	79.2100	2.98329	9.43398
8.41	70.7281	2.90000	9.17061	8.91	79.3881	2.98496	9.43928
8.42	70.8964	2.90172	9.17606	8.92	79.5664	2.98664	9.44458
8.43	71.0649	2.90345	9.18150	8.93	79.7449	2.98831	9.44987
8.44	71.2336	2.90517	9.18695	8.94	79.9236	2.98998	9.45516
8.45	71.4025	2.90689	9.19239	8.95	80.1025	2.99166	9.46044
8.46	71.5716	2.90861	9.19783	8.96	80.2816	2.99333	9.46573
8.47	71.7409	2.91033	9.20326	8.97	80.4609	2.99500	9.47101
8.48	71.9104	2.91204	9.20869	8.98	80.6404	2.99666	9.47629
8.49	72.0801	2.91376	9.21412	8.99	80.8201	2.99833	9.48156
8.50	72.2500	2.91548	9.21954	**9.00**	81.0000	3.00000	9.48683
N	N²	√N	√10N	N	N²	√N	√10N

TABLE I. (*Continued*)

N	N²	√N̄	√1̄0̄N̄		N	N²	√N̄	√1̄0̄N̄
9.00	81.0000	3.00000	9.48683		**9.50**	90.2500	3.08221	9.74679
9.01	81.1801	3.00167	9.49210		9.51	90.4401	3.08383	9.75192
9.02	81.3604	3.00333	9.49737		9.52	90.6304	3.08545	9.75705
9.03	81.5409	3.00500	9.50263		9.53	90.8209	3.08707	9.76217
9.04	81.7216	3.00666	9.50789		9.54	91.0116	3.08869	9.76729
9.05	81.9025	3.00832	9.51315		9.55	91.2025	3.09031	9.77241
9.06	82.0836	3.00998	9.51840		9.56	91.3936	3.09192	9.77753
9.07	82.2649	3.01164	9.52365		9.57	91.5849	3.09354	9.78264
9.08	82.4464	3.01330	9.52890		9.58	91.7764	3.09516	9.78775
9.09	82.6281	3.01496	9.53415		9.59	91.9681	3.09677	9.79285
9.10	82.8100	3.01662	9.53939		**9.60**	92.1600	3.09839	9.79796
9.11	82.9921	3.01828	9.54463		9.61	92.3521	3.10000	9.80306
9.12	83.1744	3.01993	9.54987		9.62	92.5444	3.10161	9.80816
9.13	83.3569	3.02159	9.55510		9.63	92.7369	3.10322	9.81326
9.14	83.5396	3.02324	9.56033		9.64	92.9296	3.10483	9.81835
9.15	83.7225	3.02490	9.56556		9.65	93.1225	3.10644	9.82344
9.16	83.9056	3.02655	9.57079		9.66	93.3156	3.10805	9.82853
9.17	84.0889	3.02820	9.57601		9.67	93.5089	3.10966	9.83362
9.18	84.2724	3.02985	9.58123		9.68	93.7024	3.11127	9.83870
9.19	84.4561	3.03150	9.58645		9.69	93.8961	3.11288	9.84378
9.20	84.6400	3.03315	9.59166		**9.70**	94.0900	3.11448	9.84886
9.21	84.8241	3.03480	9.59687		9.71	94.2841	3.11609	9.85393
9.22	85.0084	3.03645	9.60208		9.72	94.4784	3.11769	9.85901
9.23	85.1929	3.03809	9.60729		9.73	94.6729	3.11929	9.86408
9.24	85.3776	3.03974	9.61249		9.74	94.8676	3.12090	9.86914
9.25	85.5625	3.04138	9.61769		9.75	95.0625	3.12250	9.87421
9.26	85.7476	3.04302	9.62289		9.76	95.2576	3.12410	9.87927
9.27	85.9329	3.04467	9.62808		9.77	95.4529	3.12570	9.88433
9.28	86.1184	3.04631	9.63328		9.78	95.6484	3.12730	9.88939
9.29	86.3041	3.04795	9.63846		9.79	95.8441	3.12890	9.89444
9.30	86.4900	3.04959	9.64365		**9.80**	96.0400	3.13050	9.89949
9.31	86.6761	3.05123	9.64883		9.81	96.2361	3.13209	9.90454
9.32	86.8624	3.05287	9.65401		9.82	96.4324	3.13369	9.90959
9.33	87.0489	3.05450	9.65919		9.83	96.6289	3.13528	9.91464
9.34	87.2356	3.05614	9.66437		9.84	96.8256	3.13688	9.91968
9.35	87.4225	3.05778	9.66954		9.85	97.0225	3.13847	9.92472
9.36	87.6096	3.05941	9.67471		9.86	97.2196	3.14006	9.92974
9.37	87.7969	3.06105	9.67988		9.87	97.4169	3.14166	9.93479
9.38	87.9844	3.06268	9.68504		9.88	97.6144	3.14325	9.93982
9.39	88.1721	3.06431	9.69020		9.89	97.8121	3.14484	9.94485
9.40	88.3600	3.06594	9.69536		**9.90**	98.0100	3.14643	9.94987
9.41	88.5481	3.06757	9.70052		9.91	98.2081	3.14802	9.95490
9.42	88.7364	3.06920	9.70567		9.92	98.4064	3.14960	9.95992
9.43	88.9249	3.07083	9.71082		9.93	98.6049	3.15119	9.96494
9.44	89.1136	3.07246	9.71597		9.94	98.8036	3.15278	9.96995
9.45	89.3025	3.07409	9.72111		9.95	99.0025	3.15436	9.97497
9.46	89.4916	3.07571	9.72625		9.96	99.2016	3.15595	9.97998
9.47	89.6809	3.07734	9.73139		9.97	99.4009	3.15753	9.98499
9.48	89.8704	3.07896	9.73653		9.98	99.6004	3.15911	9.98999
9.49	90.0601	3.08058	9.74166		9.99	99.8001	3.16070	9.99500
9.50	90.2500	3.08221	9.74679		**10.0**	100.000	3.16228	10.0000
N	N²	√N̄	√1̄0̄N̄		N	N²	√N̄	√1̄0̄N̄

283

(computer) X = rn D

TABLE II. Random Digits

03991	10461	93716	16894	98953	73231	39528	72484	82474	25593
38555	95554	32886	59780	09958	18065	81616	18711	53342	44276
17546	73704	92052	46215	15917	06253	07586	16120	82641	22820
32643	52861	95819	06831	19640	99413	90767	04235	13574	17200
69572	68777	39510	35905	85244	35159	40188	28193	29593	88627
24122	66591	27699	06494	03152	19121	34414	82157	86887	55087
61196	30231	92962	61773	22109	78508	63439	75363	44989	16822
30532	21704	10274	12202	94205	20380	67049	09070	93399	45547
03788	97599	75867	20717	82037	10268	79495	04146	52162	90286
48228	63379	85783	47619	87481	37220	91704	30552	04737	21031
88618	19161	41290	67312	71857	15957	48545	35247	18619	13674
71299	23853	05870	01119	92784	26340	75122	11724	74627	73707
27954	58909	82444	99005	04921	73701	92904	13141	32392	19763
80863	00514	20247	81759	45197	25332	69902	63742	78464	22501
33564	60780	48460	85558	15191	18782	94972	11598	62095	36787
90899	75754	60833	25983	01291	41349	19152	00023	12302	80783
78038	70267	43529	06318	38384	74761	36024	00867	76378	41605
55986	66485	88722	56736	66164	49431	94458	74284	05041	49807
87539	08823	94813	31900	54155	83436	54158	34243	46978	35482
16818	60311	74457	90561	72848	11834	75051	93029	47665	64382
34677	58300	74910	64345	19325	81549	60365	94653	35075	33949
45305	07521	61318	31855	14413	70951	83799	42402	56623	34442
59747	67277	76503	34513	39663	77544	32960	07405	36409	83232
16520	69676	11654	99893	02181	68161	19322	53845	57620	52606
68652	27376	92852	55866	88448	03584	11220	94747	07399	37408
79375	95220	01159	63267	10622	48391	31751	57260	68980	05339
33521	26665	55823	47641	86225	31704	88492	99382	14454	04504
59589	49067	66821	41575	49767	04037	30934	47744	07481	83828
20554	91409	96277	48257	50816	97616	22888	48893	27499	98748
59404	72059	43947	51680	43852	59693	78212	16993	35902	91386
42614	29297	01918	28316	25163	01889	70014	15021	68971	11403
34994	41374	70071	14736	65251	07629	37239	33295	18477	65622
99385	41600	11133	07586	36815	43625	18637	37509	14707	93997
66497	68646	78138	66559	64397	11692	05327	82162	83745	22567
48509	23929	27482	45476	04515	25624	95096	67946	16930	33361
15470	48355	88651	22596	83761	60873	43253	84145	20368	07126
20094	98977	74843	93413	14387	06345	80854	09279	41196	37480
73788	06533	28597	20405	51321	92246	80088	77074	66919	31678
60530	45128	74022	84617	72472	00008	80890	18002	35352	54131
44372	15486	65741	14014	05466	55306	93128	18464	79982	68416
18611	19241	66083	24653	84609	58232	41849	84547	46850	52326
58319	15997	08355	60860	29735	47762	46352	33049	69248	93460
61199	67940	55121	29281	59076	07936	11087	96294	14013	31792
18627	90872	00911	98936	76355	93779	52701	08337	56303	87315
00441	58997	14060	40619	29549	69616	57275	36898	81304	48585
32624	68691	14845	46672	61958	77100	20857	73156	70284	24326
65961	73488	41839	55382	17267	70943	15633	84924	90415	93614
20288	34060	39685	23309	10061	68829	92694	48297	39904	02115
59362	95938	74416	53166	35208	33374	77613	19019	88152	00080
99782	93478	53152	67433	35663	52972	38688	32486	45134	63545

TABLE II. *(Continued)*

27767	43584	85301	88977	29490	69714	94015	64874	32444	48277
13025	14338	54066	15243	47724	66733	74108	88222	88570	74015
80217	36292	98525	24335	24432	24896	62880	87873	95160	59221
10875	62004	90391	61105	57411	06368	11748	12102	80580	41867
54127	57326	26629	19087	24472	88779	17944	05600	60478	03343
60311	42824	37301	42678	45990	43242	66067	42792	95043	52680
49739	71484	92003	98086	76668	73209	54244	91030	45547	70818
78626	51594	16453	94614	39014	97066	30945	57589	31732	57260
66692	13986	99837	00582	81232	44987	69170	37403	86995	90307
44071	28091	07362	97703	76447	42537	08345	88975	35841	85771
59820	96163	78851	16499	87064	13075	73035	41207	74699	09310
25704	91035	26313	77463	55387	72681	47431	43905	31048	56699
22304	90314	78438	66276	18396	73538	43277	58874	11466	16082
17710	59621	15292	76139	59526	52113	53856	30743	08670	84741
25852	58905	55018	56374	35824	71708	30540	27886	61732	75454
46780	56487	75211	10271	36633	68424	17374	52003	70707	70214
59849	96169	87195	46092	26787	60939	59202	11973	02902	33250
47670	07654	30342	40277	11049	72049	83012	09832	25571	77628
94304	71803	73465	09819	58869	35220	09504	96412	90193	79568
08105	59987	21437	36786	49226	77837	98524	97831	65704	09514
64281	61826	18555	64937	64654	25843	41145	42820	14924	39650
66847	70495	32350	02985	01755	14750	48968	38603	70312	05682
72461	33230	21529	53424	72877	17334	39283	04149	90850	64618
21032	91050	13058	16218	06554	07850	73950	79552	24781	89683
95362	67011	06651	16136	57216	39618	49856	99326	40902	05069
49712	97380	10404	55452	09971	59481	37006	22186	72682	07385
58275	61764	97586	54716	61459	21647	87417	17198	21443	41808
89514	11788	68224	23417	46376	25366	94746	49580	01176	28838
15472	50669	48139	36732	26825	05511	12459	91314	80582	71944
12120	86124	51247	44302	87112	21476	14713	71181	13177	55292
95294	00556	70481	06905	21785	41101	49386	54480	23604	23554
66986	34099	74474	20740	47458	64809	06312	88940	15995	69321
80620	51790	11436	38072	40405	68032	60942	00307	11897	92674
55411	85667	77535	99892	71209	92061	92329	98932	78284	46347
95083	06783	28102	57816	85561	29671	77936	63574	31384	51924
90726	57166	98884	08583	95889	57067	38101	77756	11657	13897
68984	83620	89747	98882	92613	89719	39641	69457	91339	22502
36421	16489	18059	51061	67667	60631	84054	40455	99396	63680
92638	40333	67054	16067	24700	71594	47468	03577	57649	63266
21036	82808	77501	97427	76479	68562	43321	31370	28977	23896
13173	33365	41468	85149	49554	17994	91178	10174	29420	90438
86716	38746	94559	37559	49678	53119	98189	81851	29651	84215
92581	02262	41615	70360	64114	58660	96717	54244	10701	41393
12470	56500	50273	93113	41794	86861	39448	93136	25722	08564
01016	00857	41396	80504	90670	08289	58137	17820	22751	36518
34030	60726	25807	24260	71529	78920	47648	13885	70669	93406
50259	46345	06170	97965	88302	98041	11947	56203	19324	20504
73958	76145	60808	54444	74412	81105	69181	96845	38525	11600
46874	37088	80940	44893	10408	36222	14004	23153	69249	05747
60883	52109	19516	90120	46759	71643	62342	07589	08899	05985

TABLE III. **Binomial Coefficients** $\dfrac{n!}{x!(n-x)!}$

n \ x	2	3	4	5	6	7	8	9	10
2	1								
3	3	1							
4	6	4	1						
5	10	10	5	1					
6	15	20	15	6	1				
7	21	35	35	21	7	1			
8	28	56	70	56	28	8	1		
9	36	84	126	126	84	36	9	1	
10	45	120	210	252	210	120	45	10	1
11	55	165	330	462	462	330	165	55	11
12	66	220	495	792	924	792	495	220	66
13	78	286	715	1,287	1,716	1,716	1,287	715	286
14	91	364	1,001	2,002	3,003	3,432	3,003	2,002	1,001
15	105	455	1,365	3,003	5,005	6,435	6,435	5,005	3,003
16	120	560	1,820	4,368	8,008	11,440	12,870	11,440	8,008
17	136	680	2,380	6,188	12,376	19,448	24,310	24,310	19,448
18	153	816	3,060	8,568	18,564	31,824	43,758	48,620	43,758
19	171	969	3,876	11,628	27,132	50,388	75,582	92,378	92,378
20	190	1,140	4,845	15,504	38,760	77,520	125,970	167,960	184,756

$P \leq \tfrac{1}{2} \quad pn > 5$

$P > \tfrac{1}{2} \quad qn > 5$

sample use $z_{\bar{x}} = \dfrac{\bar{x} - \mu}{\sigma_{\bar{x}}}$

sample $\sigma_{\bar{x}} = \dfrac{\sigma}{\sqrt{n}}$

$\mu_{\bar{x}} = \mu$

$x = \mu + z\sigma$

495?
495?
99(4
.09
.2|98
8
18

2.33

TABLE IV. Areas of a Standard Normal Distribution

An entry in the table is the proportion under the entire curve which is between $z = 0$ and a positive value of z. Areas for negative values of z are obtained by symmetry.

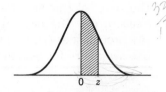

.5000
.3212
.1788

z	.00	.01	.02	.03	.04	.05	.06	.07	.08	.09
0.0	.0000	.0040	.0080	.0120	.0160	.0199	.0239	.0279	.0319	.0359
0.1	.0398	.0438	.0478	.0517	.0557	.0596	.0636	.0675	.0714	.0753
0.2	.0793	.0832	.0871	.0910	.0948	.0987	.1026	.1064	.1103	.1141
0.3	.1179	.1217	.1255	.1293	.1331	.1368	.1406	.1443	.1480	.1517
0.4	.1554	.1591	.1628	.1664	.1700	.1736	.1772	.1808	.1844	.1879
0.5	.1915	.1950	.1985	.2019	.2054	.2088	.2123	.2157	.2190	.2224
0.6	.2257	.2291	.2324	.2357	.2389	.2422	.2454	.2486	.2517	.2549
0.7	.2580	.2611	.2642	.2673	.2703	.2734	.2764	.2794	.2823	.2852
0.8	.2881	.2910	.2939	.2967	.2995	.3023	.3051	.3078	.3106	.3133
0.9	.3159	.3186	.3212	.3238	.3264	.3289	.3315	.3340	.3365	.3389
1.0	.3413	.3438	.3461	.3485	.3508	.3531	.3554	.3577	.3599	.3621
1.1	.3643	.3665	.3686	.3708	.3729	.3749	.3770	.3790	.3810	.3830
1.2	.3849	.3869	.3888	.3907	.3925	.3944	.3962	.3980	.3997	.4015
1.3	.4032	.4049	.4066	.4082	.4099	.4115	.4131	.4147	.4162	.4177
1.4	.4192	.4207	.4222	.4236	.4251	.4265	.4279	.4292	.4306	.4319
1.5	.4332	.4345	.4357	.4370	.4382	.4394	.4406	.4418	.4429	.4441
1.6	.4452	.4463	.4474	.4484	.4495	.4505	.4515	.4525	.4535	.4545
1.7	.4554	.4564	.4573	.4582	.4591	.4599	.4608	.4616	.4625	.4633
1.8	.4641	.4649	.4656	.4664	.4671	.4678	.4686	.4693	.4699	.4706
1.9	.4713	.4719	.4726	.4732	.4738	.4744	.4750	.4756	.4761	.4767
2.0	.4772	.4778	.4783	.4788	.4793	.4798	.4803	.4808	.4812	.4817
2.1	.4821	.4826	.4830	.4834	.4838	.4842	.4846	.4850	.4854	.4857
2.2	.4861	.4864	.4868	.4871	.4875	.4878	.4881	.4884	.4887	.4890
2.3	.4893	.4896	.4898	.4901	.4904	.4906	.4909	.4911	.4913	.4916
2.4	.4918	.4920	.4922	.4925	.4927	.4929	.4931	.4932	.4934	.4936
2.5	.4938	.4940	.4941	.4943	.4945	.4946	.4948	.4949	.4951	.4952
2.6	.4953	.4955	.4956	.4957	.4959	.4960	.4961	.4962	.4963	.4964
2.7	.4965	.4966	.4967	.4968	.4969	.4970	.4971	.4972	.4973	.4974
2.8	.4974	.4975	.4976	.4977	.4977	.4978	.4979	.4979	.4980	.4981
2.9	.4981	.4982	.4982	.4983	.4984	.4984	.4985	.4985	.4986	.4986
3.0	.4987	.4987	.4987	.4988	.4988	.4989	.4989	.4989	.4990	.4990

testing mean - p. 104
testing diff btwn. 2 means - 171
testing a proportion - p. 175

testing diff of 2 prop. - p. 178

TABLE V. Student's t Distribution

The first column lists the number of degrees of
freedom (ν). The headings of the other columns
give probabilities (P) for t to exceed the entry value.
Use symmetry for negative t values.

areas in top row

$r = n-1$

ν	P .10	.05	.025	.01	.005
1	3.078	6.314	12.706	31.821	63.657
2	1.886	2.920	4.303	6.965	9.925
3	1.638	2.353	3.182	4.541	5.841
4	1.533	2.132	2.776	3.747	4.604
5	1.476	2.015	2.571	3.365	4.032
6	1.440	1.943	2.447	3.143	3.707
7	1.415	1.895	2.365	2.998	3.499
8	1.397	1.860	2.306	2.896	3.355
9	1.383	1.833	2.262	2.821	3.250
10	1.372	1.812	2.228	2.764	3.169
11	1.363	1.796	2.201	2.718	3.106
12	1.356	1.782	2.179	2.681	3.055
13	1.350	1.771	2.160	2.650	3.012
14	1.345	1.761	2.145	2.624	2.977
15	1.341	1.753	2.131	2.602	2.947
16	1.337	1.746	2.120	2.583	2.921
17	1.333	1.740	2.110	2.567	2.898
18	1.330	1.734	2.101	2.552	2.878
19	1.328	1.729	2.093	2.539	2.861
20	1.325	1.725	2.086	2.528	2.845
21	1.323	1.721	2.080	2.518	2.831
22	1.321	1.717	2.074	2.508	2.819
23	1.319	1.714	2.069	2.500	2.807
24	1.318	1.711	2.064	2.492	2.797
25	1.316	1.708	2.060	2.485	2.787
26	1.315	1.706	2.056	2.479	2.779
27	1.314	1.703	2.052	2.473	2.771
28	1.313	1.701	2.048	2.467	2.763
29	1.311	1.699	2.045	2.462	2.756
30	1.310	1.697	2.042	2.457	2.750
40	1.303	1.684	2.021	2.423	2.704
60	1.296	1.671	2.000	2.390	2.660
120	1.289	1.658	1.980	2.358	2.617
∞	1.282	1.645	1.960	2.326	2.576

TABLE VI. Critical values of *r* for testing $\rho = 0$

For a two-sided test, α is twice the value listed at the heading of a column of critical *r* values, hence for $\alpha = .05$ choose the .025 column.

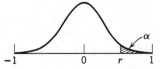

n $\diagdown$ α	.05	.025	.005
5	.805	.878	.959
6	.729	.811	.917
7	.669	.754	.875
8	.621	.707	.834
9	.582	.666	.798
10	.549	.632	.765
11	.521	.602	.735
12	.497	.576	.708
13	.476	.553	.684
14	.457	.532	.661
15	441	.514	.641
16	.426	.497	.623

n $\diagdown$ α	.05	.025	.005
17	.412	.482	.606
18	.400	.468	.590
19	.389	.456	.575
20	.378	.444	.561
25	.337	.396	.505
30	.306	.361	.463
35	.283	.334	.430
40	.264	.312	.402
50	.235	.279	.361
60	.214	.254	.330
80	.185	.220	.286
100	.165	.196	.256

TABLE VII. The χ^2 Distribution

The first column lists the number of degrees of freedom (ν). The headings of the other columns give probabilities (P) for χ^2 to exceed the entry value. For $\nu >$ 100, treat $\sqrt{2\chi^2} - \sqrt{2\nu - 1}$ as a standard normal variable.

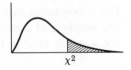

ν \\ P	0.995	0.975	0.050	0.025	0.010	0.005
1	0.0⁴3927	0.0³9821	3.84146	5.02389	6.63490	7.87944
2	0.010025	0.050636	5.99147	7.37776	9.21034	10.5966
3	0.071721	0.215795	7.81473	9.34840	11.3449	12.8381
4	0.206990	0.484419	9.48773	11.1433	13.2767	14.8602
5	0.411740	0.831211	11.0705	12.8325	15.0863	16.7496
6	0.675727	1.237347	12.5916	14.4494	16.8119	18.5476
7	0.989265	1.68987	14.0671	16.0128	18.4753	20.2777
8	1.344419	2.17973	15.5073	17.5346	20.0902	21.9550
9	1.734926	2.70039	16.9190	19.0228	21.6660	23.5893
10	2.15585	3.24697	18.3070	20.4831	23.2093	25.1882
11	2.60321	3.81575	19.6751	21.9200	24.7250	26.7569
12	3.07382	4.40379	21.0261	23.3367	26.2170	28.2995
13	3.56503	5.00874	22.3621	24.7356	27.6883	29.8194
14	4.07468	5.62872	23.6848	26.1190	29.1413	31.3193
15	4.60094	6.26214	24.9958	27.4884	30.5779	32.8013
16	5.14224	6.90766	26.2962	28.8454	31.9999	34.2672
17	5.69724	7.56418	27.5871	30.1910	33.4087	35.7185
18	6.26481	8.23075	28.8693	31.5264	34.8053	37.1564
19	6.84398	8.90655	30.1435	32.8523	36.1908	38.5822
20	7.43386	9.59083	31.4104	34.1696	37.5662	39.9968
21	8.03366	10.28293	32.6705	35.4789	38.9321	41.4010
22	8.64272	10.9823	33.9244	36.7807	40.2894	42.7956
23	9.26042	11.6885	35.1725	38.0757	41.6384	44.1813
24	9.88623	12.4001	36.4151	39.3641	42.9798	45.5585
25	10.5197	13.1197	37.6525	40.6465	44.3141	46.9278
26	11.1603	13.8439	38.8852	41.9232	45.6417	48.2899
27	11.8076	14.5733	40.1133	43.1944	46.9630	49.6449
28	12.4613	15.3079	41.3372	44.4607	48.2782	50.9933
29	13.1211	16.0471	42.5569	45.7222	49.5879	52.3356
30	13.7867	16.7908	43.7729	46.9792	50.8922	53.6720
40	20.7065	24.4331	55.7585	59.3417	63.6907	66.7659
50	27.9907	32.3574	67.5048	71.4202	76.1539	79.4900
60	35.5346	40.4817	79.0819	83.2976	88.3794	91.9517
70	43.2752	48.7576	90.5312	95.0231	100.425	104.215
80	51.1720	57.1532	101.879	106.629	112.329	116.321
90	59.1963	65.6466	113.145	118.136	124.116	128.299
100	67.3276	74.2219	124.342	129.561	135.807	140.169

TABLE VIII. Rank-Sum Critical Values

The sample sizes are shown in parentheses (n_1, n_2). The probability associated with a pair of critical values is the probability that $R \leq$ smaller value, or equally, it is the probability that $R \geq$ larger value. These probabilities are the closest ones to .025 and .05 that exist for integer values of R. The approximate .025 values should be used for a two-sided test with $\alpha = .05$, and the approximate .05 values for a one-sided test.

(2, 4)			(4, 4)			(6, 7)		
3	11	.067	11	25	.029	28	56	.026
(2, 5)			12	24	.057	30	54	.051
3	13	.047	(4, 5)			(6, 8)		
(2, 6)			12	28	.032	29	61	.021
3	15	.036	13	27	.056	32	58	.054
4	14	.071	(4, 6)			(6, 9)		
(2, 7)			12	32	.019	31	65	.025
3	17	.028	14	30	.057	33	63	.044
4	16	.056	(4, 7)			(6, 10)		
(2, 8)			13	35	.021	33	69	.028
3	19	.022	15	33	.055	35	67	.047
4	18	.044	(4, 8)			(7, 7)		
(2, 9)			14	38	.024	37	68	.027
3	21	.018	16	36	.055	39	66	.049
4	20	.036	(4, 9)			(7, 8)		
(2, 10)			15	41	.025	39	73	.027
4	22	.030	17	39	.053	41	71	.047
5	21	.061	(4, 10)			(7, 9)		
(3, 3)			16	44	.026	41	78	.027
6	15	.050	18	42	.053	43	76	.045
(3, 4)			(5, 5)			(7, 10)		
6	18	.028	18	37	.028	43	83	.028
7	17	.057	19	36	.048	46	80	.054
(3, 5)			(5, 6)			(8, 8)		
6	21	.018	19	41	.026	49	87	.025
7	20	.036	20	40	.041	52	84	.052
(3, 6)			(5, 7)			(8, 9)		
7	23	.024	20	45	.024	51	93	.023
8	22	.048	22	43	.053	54	90	.046
(3, 7)			(5, 8)			(8, 10)		
8	25	.033	21	49	.023	54	98	.027
9	24	.058	23	47	.047	57	95	.051
(3, 8)			(5, 9)			(9, 9)		
8	28	.024	22	53	.021	63	108	.025
9	27	.042	25	50	.056	66	105	.047
(3, 9)			(5, 10)			(9, 10)		
9	30	.032	24	56	.028	66	114	.027
10	29	.050	26	54	.050	69	111	.047
(3, 10)			(6, 6)			(10, 10)		
9	33	.024	26	52	.021	79	131	.026
11	31	.056	28	50	.047	83	127	.053

TABLE IX.   F Distribution

5% (Roman Type) and 1% (Boldface Type) Points for the Distribution of F

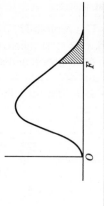

Degrees of freedom for numerator (v_1)

Degrees of freedom for denominator (v_2)	1	2	3	4	5	6	7	8	9	10	11	12	14	16	20	24	30	40	50	75	100	200	500	∞
1	161 **4052**	200 **4999**	216 **5403**	225 **5625**	230 **5764**	234 **5859**	237 **5928**	239 **5981**	241 **6022**	242 **6056**	243 **6082**	244 **6106**	245 **6142**	246 **6169**	248 **6208**	249 **6234**	250 **6258**	251 **6286**	252 **6302**	253 **6323**	253 **6334**	254 **6352**	254 **6361**	254 **6366**
2	18.51 **98.49**	19.00 **99.01**	19.16 **99.17**	19.25 **99.25**	19.30 **99.30**	19.33 **99.33**	19.36 **99.34**	19.37 **99.36**	19.38 **99.38**	19.39 **99.40**	19.40 **99.41**	19.41 **99.42**	19.42 **99.43**	19.43 **99.44**	19.44 **99.45**	19.45 **99.46**	19.46 **99.47**	19.47 **99.48**	19.47 **99.48**	19.48 **99.49**	19.49 **99.49**	19.49 **99.49**	19.50 **99.50**	19.50 **99.50**
3	10.13 **34.12**	9.55 **30.81**	9.28 **29.46**	9.12 **28.71**	9.01 **28.24**	8.94 **27.91**	8.88 **27.67**	8.84 **27.49**	8.81 **27.34**	8.78 **27.23**	8.76 **27.13**	8.74 **27.05**	8.71 **26.92**	8.69 **26.83**	8.66 **26.69**	8.64 **26.60**	8.62 **26.50**	8.60 **26.41**	8.58 **26.30**	8.57 **26.27**	8.56 **26.23**	8.54 **26.18**	8.54 **26.14**	8.53 **26.12**
4	7.71 **21.20**	6.94 **18.00**	6.59 **16.69**	6.39 **15.98**	6.26 **15.52**	6.16 **15.21**	6.09 **14.98**	6.04 **14.80**	6.00 **14.66**	5.96 **14.54**	5.93 **14.45**	5.91 **14.37**	5.87 **14.24**	5.84 **14.15**	5.80 **14.02**	5.77 **13.93**	5.74 **13.83**	5.71 **13.74**	5.70 **13.69**	5.68 **13.61**	5.66 **13.57**	5.65 **13.52**	5.64 **13.48**	5.63 **13.46**
5	6.61 **16.26**	5.79 **13.27**	5.41 **12.06**	5.19 **11.39**	5.05 **10.97**	4.95 **10.67**	4.88 **10.45**	4.82 **10.27**	4.78 **10.15**	4.74 **10.05**	4.70 **9.96**	4.68 **9.89**	4.64 **9.77**	4.60 **9.68**	4.56 **9.55**	4.53 **9.47**	4.50 **9.38**	4.46 **9.29**	4.44 **9.24**	4.42 **9.17**	4.40 **9.13**	4.38 **9.07**	4.37 **9.04**	4.36 **9.02**
6	5.99 **13.74**	5.14 **10.92**	4.76 **9.78**	4.53 **9.15**	4.39 **8.75**	4.28 **8.47**	4.21 **8.26**	4.15 **8.10**	4.10 **7.98**	4.06 **7.87**	4.03 **7.79**	4.00 **7.72**	3.96 **7.60**	3.92 **7.52**	3.87 **7.39**	3.84 **7.31**	3.81 **7.23**	3.77 **7.14**	3.75 **7.09**	3.72 **7.02**	3.71 **6.99**	3.69 **6.94**	3.68 **6.90**	3.67 **6.88**
7	5.59 **12.25**	4.74 **9.55**	4.35 **8.45**	4.12 **7.85**	3.97 **7.46**	3.87 **7.19**	3.79 **7.00**	3.73 **6.84**	3.68 **6.71**	3.63 **6.62**	3.60 **6.54**	3.57 **6.47**	3.52 **6.35**	3.49 **6.27**	3.44 **6.15**	3.41 **6.07**	3.38 **5.98**	3.34 **5.90**	3.32 **5.85**	3.29 **5.78**	3.28 **5.75**	3.25 **5.70**	3.24 **5.67**	3.23 **5.65**
8	5.32 **11.26**	4.46 **8.65**	4.07 **7.59**	3.84 **7.01**	3.69 **6.63**	3.58 **6.37**	3.50 **6.19**	3.44 **6.03**	3.39 **5.91**	3.34 **5.82**	3.31 **5.74**	3.28 **5.67**	3.23 **5.56**	3.20 **5.48**	3.15 **5.36**	3.12 **5.28**	3.08 **5.20**	3.05 **5.11**	3.03 **5.06**	3.00 **5.00**	2.98 **4.96**	2.96 **4.91**	2.94 **4.88**	2.93 **4.86**
9	5.12 **10.56**	4.26 **8.02**	3.86 **6.99**	3.63 **6.42**	3.48 **6.06**	3.37 **5.80**	3.29 **5.62**	3.23 **5.47**	3.18 **5.35**	3.13 **5.26**	3.10 **5.18**	3.07 **5.11**	3.02 **5.00**	2.98 **4.92**	2.93 **4.80**	2.90 **4.73**	2.86 **4.64**	2.82 **4.56**	2.80 **4.51**	2.77 **4.45**	2.76 **4.41**	2.73 **4.36**	2.72 **4.33**	2.71 **4.31**

10	2.54 / 3.91	2.55 / 3.93	2.56 / 3.96	2.59 / 4.01	2.61 / 4.05	2.64 / 4.12	2.67 / 4.17	2.70 / 4.25	2.74 / 4.33	2.77 / 4.41	2.82 / 4.52	2.86 / 4.60	2.91 / 4.71	2.94 / 4.78	2.97 / 4.85	3.02 / 4.95	3.07 / 5.06	3.14 / 5.21	3.22 / 5.39	3.33 / 5.64	3.48 / 5.99	3.71 / 6.55	4.10 / 7.56	4.96 / 10.04
11	2.40 / 3.60	2.41 / 3.62	2.42 / 3.66	2.45 / 3.70	2.47 / 3.74	2.50 / 3.80	2.53 / 3.86	2.57 / 3.94	2.61 / 4.02	2.65 / 4.10	2.70 / 4.21	2.74 / 4.29	2.79 / 4.40	2.82 / 4.46	2.86 / 4.54	2.90 / 4.63	2.95 / 4.74	3.01 / 4.88	3.09 / 5.07	3.20 / 5.32	3.36 / 5.67	3.59 / 6.22	3.98 / 7.20	4.84 / 9.65
12	2.30 / 3.36	2.31 / 3.38	2.32 / 3.41	2.35 / 3.46	2.36 / 3.49	2.40 / 3.56	2.42 / 3.61	2.46 / 3.70	2.50 / 3.78	2.54 / 3.86	2.60 / 3.98	2.64 / 4.05	2.69 / 4.16	2.72 / 4.22	2.76 / 4.30	2.80 / 4.39	2.85 / 4.50	2.92 / 4.65	3.00 / 4.82	3.11 / 5.06	3.26 / 5.41	3.49 / 5.95	3.88 / 6.93	4.75 / 9.33
13	2.21 / 3.16	2.22 / 3.18	2.24 / 3.21	2.26 / 3.27	2.28 / 3.30	2.32 / 3.37	2.34 / 3.42	2.38 / 3.51	2.42 / 3.59	2.46 / 3.67	2.51 / 3.78	2.55 / 3.85	2.60 / 3.96	2.63 / 4.02	2.67 / 4.10	2.72 / 4.19	2.77 / 4.30	2.84 / 4.44	2.92 / 4.62	3.02 / 4.86	3.18 / 5.20	3.41 / 5.74	3.80 / 6.70	4.67 / 9.07
14	2.13 / 3.00	2.14 / 3.02	2.16 / 3.06	2.19 / 3.11	2.21 / 3.14	2.24 / 3.21	2.27 / 3.26	2.31 / 3.34	2.35 / 3.43	2.39 / 3.51	2.44 / 3.62	2.48 / 3.70	2.53 / 3.80	2.56 / 3.86	2.60 / 3.94	2.65 / 4.03	2.70 / 4.14	2.77 / 4.28	2.85 / 4.46	2.96 / 4.69	3.11 / 5.03	3.34 / 5.56	3.74 / 6.51	4.60 / 8.86
15	2.07 / 2.87	2.08 / 2.89	2.10 / 2.92	2.12 / 2.97	2.15 / 3.00	2.18 / 3.07	2.21 / 3.12	2.25 / 3.20	2.29 / 3.29	2.33 / 3.36	2.39 / 3.48	2.43 / 3.56	2.48 / 3.67	2.51 / 3.73	2.55 / 3.80	2.59 / 3.89	2.64 / 4.00	2.70 / 4.14	2.79 / 4.32	2.90 / 4.56	3.06 / 4.89	3.29 / 5.42	3.68 / 6.36	4.54 / 8.68
16	2.01 / 2.75	2.02 / 2.77	2.04 / 2.80	2.07 / 2.86	2.09 / 2.89	2.13 / 2.96	2.16 / 3.01	2.20 / 3.10	2.24 / 3.18	2.28 / 3.25	2.33 / 3.37	2.37 / 3.45	2.42 / 3.55	2.45 / 3.61	2.49 / 3.69	2.54 / 3.78	2.59 / 3.89	2.66 / 4.03	2.74 / 4.20	2.85 / 4.44	3.01 / 4.77	3.24 / 5.29	3.63 / 6.23	4.49 / 8.53
17	1.96 / 2.65	1.97 / 2.67	1.99 / 2.70	2.02 / 2.76	2.04 / 2.79	2.08 / 2.86	2.11 / 2.92	2.15 / 3.00	2.19 / 3.08	2.23 / 3.16	2.29 / 3.27	2.33 / 3.35	2.38 / 3.45	2.41 / 3.52	2.45 / 3.59	2.50 / 3.68	2.55 / 3.79	2.62 / 3.93	2.70 / 4.10	2.81 / 4.34	2.96 / 4.67	3.20 / 5.18	3.59 / 6.11	4.45 / 8.40
18	1.92 / 2.57	1.93 / 2.59	1.95 / 2.62	1.98 / 2.68	2.00 / 2.71	2.04 / 2.78	2.07 / 2.83	2.11 / 2.91	2.15 / 3.00	2.19 / 3.07	2.25 / 3.19	2.29 / 3.27	2.34 / 3.37	2.37 / 3.44	2.41 / 3.51	2.46 / 3.60	2.51 / 3.71	2.58 / 3.85	2.66 / 4.01	2.77 / 4.25	2.93 / 4.58	3.16 / 5.09	3.55 / 6.01	4.41 / 8.28
19	1.88 / 2.49	1.90 / 2.51	1.91 / 2.54	1.94 / 2.60	1.96 / 2.63	2.00 / 2.70	2.02 / 2.76	2.07 / 2.84	2.11 / 2.92	2.15 / 3.00	2.21 / 3.12	2.26 / 3.19	2.31 / 3.30	2.34 / 3.36	2.38 / 3.43	2.43 / 3.52	2.48 / 3.63	2.55 / 3.77	2.63 / 3.94	2.74 / 4.17	2.90 / 4.50	3.13 / 5.01	3.52 / 5.93	4.38 / 8.18
20	1.84 / 2.42	1.85 / 2.44	1.87 / 2.47	1.90 / 2.53	1.92 / 2.56	1.96 / 2.63	1.99 / 2.69	2.04 / 2.77	2.08 / 2.86	2.12 / 2.94	2.18 / 3.05	2.23 / 3.13	2.28 / 3.23	2.31 / 3.30	2.35 / 3.37	2.40 / 3.45	2.45 / 3.56	2.52 / 3.71	2.60 / 3.87	2.71 / 4.10	2.87 / 4.43	3.10 / 4.94	3.49 / 5.85	4.35 / 8.10
21	1.81 / 2.36	1.82 / 2.38	1.84 / 2.42	1.87 / 2.47	1.89 / 2.51	1.93 / 2.58	1.96 / 2.63	2.00 / 2.72	2.05 / 2.80	2.09 / 2.88	2.15 / 2.99	2.20 / 3.07	2.25 / 3.17	2.28 / 3.24	2.32 / 3.31	2.37 / 3.40	2.42 / 3.51	2.49 / 3.65	2.57 / 3.81	2.68 / 4.04	2.84 / 4.37	3.07 / 4.87	3.47 / 5.78	4.32 / 8.02
22	1.78 / 2.31	1.80 / 2.33	1.81 / 2.37	1.84 / 2.42	1.87 / 2.46	1.91 / 2.53	1.93 / 2.58	1.98 / 2.67	2.03 / 2.75	2.07 / 2.83	2.13 / 2.94	2.18 / 3.02	2.23 / 3.12	2.26 / 3.18	2.30 / 3.26	2.35 / 3.35	2.40 / 3.45	2.47 / 3.59	2.55 / 3.76	2.66 / 3.99	2.82 / 4.31	3.05 / 4.82	3.44 / 5.72	4.30 / 7.94
23	1.76 / 2.26	1.77 / 2.28	1.79 / 2.32	1.82 / 2.37	1.84 / 2.41	1.88 / 2.48·	1.91 / 2.53	1.96 / 2.62	2.00 / 2.70	2.04 / 2.78	2.10 / 2.89	2.14 / 2.97	2.20 / 3.07	2.24 / 3.14	2.28 / 3.21	2.32 / 3.30	2.38 / 3.41	2.45 / 3.54	2.53 / 3.71	2.64 / 3.94	2.80 / 4.26	3.03 / 4.76	3.42 / 5.66	4.28 / 7.88
24	1.73 / 2.21	1.74 / 2.23	1.76 / 2.27	1.80 / 2.33	1.82 / 2.36	1.86 / 2.44	1.89 / 2.49	1.94 / 2.58	1.98 / 2.66	2.02 / 2.74	2.09 / 2.85	2.13 / 2.93	2.18 / 3.03	2.22 / 3.09	2.26 / 3.17	2.30 / 3.25	2.36 / 3.36	2.43 / 3.50	2.51 / 3.67	2.62 / 3.90	2.78 / 4.22	3.01 / 4.72	3.40 / 5.61	4.26 / 7.82
25	1.71 / 2.17	1.72 / 2.19	1.74 / 2.23	1.77 / 2.29	1.80 / 2.32	1.84 / 2.40	1.87 / 2.45	1.92 / 2.54	1.96 / 2.62	2.00 / 2.70	2.06 / 2.81	2.11 / 2.89	2.16 / 2.99	2.20 / 3.05	2.24 / 3.13	2.28 / 3.21	2.34 / 3.32	2.41 / 3.46	2.49 / 3.63	2.60 / 3.86	2.76 / 4.18	2.99 / 4.68	3.38 / 5.57	4.24 / 7.77

TABLE IX. (Continued)

Degrees of freedom for numerator (ν_1)

Each cell lists the upper value (top) and lower value (bottom).

Degrees of freedom for denominator (ν_2)	1	2	3	4	5	6	7	8	9	10	11	12	14	16	20	24	30	40	50	75	100	200	500	∞
26	4.22 / 7.72	3.37 / 5.53	2.98 / 4.64	2.74 / 4.14	2.59 / 3.82	2.47 / 3.59	2.39 / 3.42	2.32 / 3.29	2.27 / 3.17	2.22 / 3.09	2.18 / 3.02	2.15 / 2.96	2.10 / 2.86	2.05 / 2.77	1.99 / 2.66	1.95 / 2.58	1.90 / 2.50	1.85 / 2.41	1.82 / 2.36	1.78 / 2.28	1.76 / 2.25	1.72 / 2.19	1.70 / 2.15	1.69 / 2.13
27	4.21 / 7.68	3.35 / 5.49	2.96 / 4.60	2.73 / 4.11	2.57 / 3.79	2.46 / 3.56	2.37 / 3.39	2.30 / 3.26	2.25 / 3.14	2.20 / 3.06	2.16 / 2.98	2.13 / 2.93	2.08 / 2.83	2.03 / 2.74	1.97 / 2.63	1.93 / 2.55	1.88 / 2.47	1.84 / 2.38	1.80 / 2.33	1.76 / 2.25	1.74 / 2.21	1.71 / 2.16	1.68 / 2.12	1.67 / 2.10
28	4.20 / 7.64	3.34 / 5.45	2.95 / 4.57	2.71 / 4.07	2.56 / 3.76	2.44 / 3.53	2.36 / 3.36	2.29 / 3.23	2.24 / 3.11	2.19 / 3.03	2.15 / 2.95	2.12 / 2.90	2.06 / 2.80	2.02 / 2.71	1.96 / 2.60	1.91 / 2.52	1.87 / 2.44	1.81 / 2.35	1.78 / 2.30	1.75 / 2.22	1.72 / 2.18	1.69 / 2.13	1.67 / 2.09	1.65 / 2.06
29	4.18 / 7.60	3.33 / 5.42	2.93 / 4.54	2.70 / 4.04	2.54 / 3.73	2.43 / 3.50	2.35 / 3.33	2.28 / 3.20	2.22 / 3.08	2.18 / 3.00	2.14 / 2.92	2.10 / 2.87	2.05 / 2.77	2.00 / 2.68	1.94 / 2.57	1.90 / 2.49	1.85 / 2.41	1.80 / 2.32	1.77 / 2.27	1.73 / 2.19	1.71 / 2.15	1.68 / 2.10	1.65 / 2.06	1.64 / 2.03
30	4.17 / 7.56	3.32 / 5.39	2.92 / 4.51	2.69 / 4.02	2.53 / 3.70	2.42 / 3.47	2.34 / 3.30	2.27 / 3.17	2.21 / 3.06	2.16 / 2.98	2.12 / 2.90	2.09 / 2.84	2.04 / 2.74	1.99 / 2.66	1.93 / 2.55	1.89 / 2.47	1.84 / 2.38	1.79 / 2.29	1.76 / 2.24	1.72 / 2.16	1.69 / 2.13	1.66 / 2.07	1.64 / 2.03	1.62 / 2.01
32	4.15 / 7.50	3.30 / 5.34	2.90 / 4.46	2.67 / 3.97	2.51 / 3.66	2.40 / 3.42	2.32 / 3.25	2.25 / 3.12	2.19 / 3.01	2.14 / 2.94	2.10 / 2.86	2.07 / 2.80	2.02 / 2.70	1.97 / 2.62	1.91 / 2.51	1.86 / 2.42	1.82 / 2.34	1.76 / 2.25	1.74 / 2.20	1.69 / 2.12	1.67 / 2.08	1.64 / 2.02	1.61 / 1.98	1.59 / 1.96
34	4.13 / 7.44	3.28 / 5.29	2.88 / 4.42	2.65 / 3.93	2.49 / 3.61	2.38 / 3.38	2.30 / 3.21	2.23 / 3.08	2.17 / 2.97	2.12 / 2.89	2.08 / 2.82	2.05 / 2.76	2.00 / 2.66	1.95 / 2.58	1.89 / 2.47	1.84 / 2.38	1.80 / 2.30	1.74 / 2.21	1.71 / 2.15	1.67 / 2.08	1.64 / 2.04	1.61 / 1.98	1.59 / 1.94	1.57 / 1.91
36	4.11 / 7.39	3.26 / 5.25	2.86 / 4.38	2.63 / 3.89	2.48 / 3.58	2.36 / 3.35	2.28 / 3.18	2.21 / 3.04	2.15 / 2.94	2.10 / 2.86	2.06 / 2.78	2.03 / 2.72	1.98 / 2.62	1.93 / 2.54	1.87 / 2.43	1.82 / 2.35	1.78 / 2.26	1.72 / 2.17	1.69 / 2.12	1.65 / 2.04	1.62 / 2.00	1.59 / 1.94	1.56 / 1.90	1.55 / 1.87
38	4.10 / 7.35	3.25 / 5.21	2.85 / 4.34	2.62 / 3.86	2.46 / 3.54	2.35 / 3.32	2.26 / 3.15	2.19 / 3.02	2.14 / 2.91	2.09 / 2.82	2.05 / 2.75	2.02 / 2.69	1.96 / 2.59	1.92 / 2.51	1.85 / 2.40	1.80 / 2.32	1.76 / 2.22	1.71 / 2.14	1.67 / 2.08	1.63 / 2.00	1.60 / 1.97	1.57 / 1.90	1.54 / 1.86	1.53 / 1.84
40	4.08 / 7.31	3.23 / 5.18	2.84 / 4.31	2.61 / 3.83	2.45 / 3.51	2.34 / 3.29	2.25 / 3.12	2.18 / 2.99	2.12 / 2.88	2.07 / 2.80	2.04 / 2.73	2.00 / 2.66	1.95 / 2.56	1.90 / 2.49	1.84 / 2.37	1.79 / 2.29	1.74 / 2.20	1.69 / 2.11	1.66 / 2.05	1.61 / 1.97	1.59 / 1.94	1.55 / 1.88	1.53 / 1.84	1.51 / 1.81
42	4.07 / 7.27	3.22 / 5.15	2.83 / 4.29	2.59 / 3.80	2.44 / 3.49	2.32 / 3.26	2.24 / 3.10	2.17 / 2.96	2.11 / 2.86	2.06 / 2.77	2.02 / 2.70	1.99 / 2.64	1.94 / 2.54	1.89 / 2.46	1.82 / 2.35	1.78 / 2.26	1.73 / 2.17	1.68 / 2.08	1.64 / 2.02	1.60 / 1.94	1.57 / 1.91	1.54 / 1.85	1.51 / 1.80	1.49 / 1.78
44	4.06 / 7.24	3.21 / 5.12	2.82 / 4.26	2.58 / 3.78	2.43 / 3.46	2.31 / 3.24	2.23 / 3.07	2.16 / 2.94	2.10 / 2.84	2.05 / 2.75	2.01 / 2.68	1.98 / 2.62	1.92 / 2.52	1.88 / 2.44	1.81 / 2.32	1.76 / 2.24	1.72 / 2.15	1.66 / 2.06	1.63 / 2.00	1.58 / 1.92	1.56 / 1.88	1.52 / 1.82	1.50 / 1.78	1.48 / 1.75
46	4.05 / 7.21	3.20 / 5.10	2.81 / 4.24	2.57 / 3.76	2.42 / 3.44	2.30 / 3.22	2.22 / 3.05	2.14 / 2.92	2.09 / 2.82	2.04 / 2.73	2.00 / 2.66	1.97 / 2.60	1.91 / 2.50	1.87 / 2.42	1.80 / 2.30	1.75 / 2.22	1.71 / 2.13	1.65 / 2.04	1.62 / 1.98	1.57 / 1.90	1.54 / 1.86	1.51 / 1.80	1.48 / 1.76	1.46 / 1.72
48	4.04 / 7.19	3.19 / 5.08	2.80 / 4.22	2.56 / 3.74	2.41 / 3.42	2.30 / 3.20	2.21 / 3.04	2.14 / 2.90	2.08 / 2.80	2.03 / 2.71	1.99 / 2.64	1.96 / 2.58	1.90 / 2.48	1.86 / 2.40	1.79 / 2.28	1.74 / 2.20	1.70 / 2.11	1.64 / 2.02	1.61 / 1.96	1.56 / 1.88	1.53 / 1.84	1.50 / 1.78	1.47 / 1.73	1.45 / 1.70

50	1.44 **1.68**	1.46 **1.71**	1.48 **1.76**	1.52 **1.82**	1.55 **1.86**	1.60 **1.94**	1.63 **2.00**	1.69 **2.10**	1.74 **2.18**	1.78 **2.26**	1.85 **2.39**	1.90 **2.46**	1.95 **2.56**	1.98 **2.62**	2.02 **2.70**	2.07 **2.78**	2.13 **2.88**	2.20 **3.02**	2.29 **3.18**	2.40 **3.41**	2.56 **3.72**	2.79 **4.20**	3.18 **5.06**	4.03 **7.17**
55	1.41 **1.64**	1.43 **1.66**	1.46 **1.71**	1.50 **1.78**	1.52 **1.82**	1.58 **1.90**	1.61 **1.96**	1.67 **2.06**	1.72 **2.15**	1.76 **2.23**	1.83 **2.35**	1.88 **2.43**	1.93 **2.53**	1.97 **2.59**	2.00 **2.66**	2.05 **2.75**	2.11 **2.85**	2.18 **2.98**	2.27 **3.15**	2.38 **3.37**	2.54 **3.68**	2.78 **4.16**	3.17 **5.01**	4.02 **7.12**
60	1.39 **1.60**	1.41 **1.63**	1.44 **1.68**	1.48 **1.74**	1.50 **1.79**	1.56 **1.87**	1.59 **1.93**	1.65 **2.03**	1.70 **2.12**	1.75 **2.20**	1.81 **2.32**	1.86 **2.40**	1.92 **2.50**	1.95 **2.56**	1.99 **2.63**	2.04 **2.72**	2.10 **2.82**	2.17 **2.95**	2.25 **3.12**	2.37 **3.34**	2.52 **3.65**	2.76 **4.13**	3.15 **4.98**	4.00 **7.08**
65	1.37 **1.56**	1.39 **1.60**	1.42 **1.64**	1.46 **1.71**	1.49 **1.76**	1.54 **1.84**	1.57 **1.90**	1.63 **2.00**	1.68 **2.09**	1.73 **2.18**	1.80 **2.30**	1.85 **2.37**	1.90 **2.47**	1.94 **2.54**	1.98 **2.61**	2.02 **2.70**	2.08 **2.79**	2.15 **2.93**	2.24 **3.09**	2.36 **3.31**	2.51 **3.62**	2.75 **4.10**	3.14 **4.95**	3.99 **7.04**
70	1.35 **1.53**	1.37 **1.56**	1.40 **1.63**	1.45 **1.69**	1.47 **1.74**	1.53 **1.82**	1.56 **1.88**	1.62 **1.98**	1.67 **2.07**	1.72 **2.15**	1.79 **2.28**	1.84 **2.35**	1.89 **2.45**	1.93 **2.51**	1.97 **2.59**	2.01 **2.67**	2.07 **2.77**	2.14 **2.91**	2.23 **3.06**	2.35 **3.29**	2.50 **3.60**	2.74 **4.08**	3.13 **4.92**	3.98 **7.01**
80	1.32 **1.49**	1.35 **1.52**	1.38 **1.57**	1.42 **1.65**	1.45 **1.70**	1.51 **1.78**	1.54 **1.84**	1.60 **1.94**	1.65 **2.03**	1.70 **2.11**	1.77 **2.24**	1.82 **2.32**	1.88 **2.41**	1.91 **2.48**	1.95 **2.55**	1.99 **2.64**	2.05 **2.74**	2.12 **2.87**	2.21 **3.04**	2.33 **3.25**	2.48 **3.56**	2.72 **4.04**	3.11 **4.88**	3.96 **6.96**
100	1.28 **1.43**	1.30 **1.46**	1.34 **1.51**	1.39 **1.59**	1.42 **1.64**	1.48 **1.73**	1.51 **1.79**	1.57 **1.89**	1.63 **1.98**	1.68 **2.06**	1.75 **2.19**	1.79 **2.26**	1.85 **2.36**	1.88 **2.43**	1.92 **2.51**	1.97 **2.59**	2.03 **2.69**	2.10 **2.82**	2.19 **2.99**	2.30 **3.20**	2.46 **3.51**	2.70 **3.98**	3.09 **4.82**	3.94 **6.90**
125	1.25 **1.37**	1.27 **1.40**	1.31 **1.46**	1.36 **1.54**	1.39 **1.59**	1.45 **1.68**	1.49 **1.75**	1.55 **1.85**	1.60 **1.94**	1.65 **2.03**	1.72 **2.15**	1.77 **2.23**	1.83 **2.33**	1.86 **2.40**	1.90 **2.47**	1.95 **2.56**	2.01 **2.65**	2.08 **2.79**	2.17 **2.95**	2.29 **3.17**	2.44 **3.47**	2.68 **3.94**	3.07 **4.78**	3.92 **6.84**
150	1.22 **1.33**	1.25 **1.37**	1.29 **1.43**	1.34 **1.51**	1.37 **1.56**	1.44 **1.66**	1.47 **1.72**	1.54 **1.83**	1.59 **1.91**	1.64 **2.00**	1.71 **2.12**	1.76 **2.20**	1.82 **2.30**	1.85 **2.37**	1.89 **2.44**	1.94 **2.53**	2.00 **2.62**	2.07 **2.76**	2.16 **2.92**	2.27 **3.13**	2.43 **3.44**	2.67 **3.91**	3.06 **4.75**	3.91 **6.81**
200	1.19 **1.28**	1.22 **1.33**	1.26 **1.39**	1.32 **1.48**	1.35 **1.53**	1.42 **1.62**	1.45 **1.69**	1.52 **1.79**	1.57 **1.88**	1.62 **1.97**	1.69 **2.09**	1.74 **2.17**	1.80 **2.28**	1.83 **2.34**	1.87 **2.41**	1.92 **2.50**	1.98 **2.60**	2.05 **2.73**	2.14 **2.90**	2.26 **3.11**	2.41 **3.41**	2.65 **3.88**	3.04 **4.71**	3.89 **6.76**
400	1.13 **1.19**	1.16 **1.24**	1.22 **1.32**	1.28 **1.42**	1.32 **1.47**	1.38 **1.57**	1.42 **1.64**	1.49 **1.74**	1.54 **1.84**	1.60 **1.92**	1.67 **2.04**	1.72 **2.12**	1.78 **2.23**	1.81 **2.29**	1.85 **2.37**	1.90 **2.46**	1.96 **2.55**	2.03 **2.69**	2.12 **2.85**	2.23 **3.06**	2.39 **3.36**	2.62 **3.83**	3.02 **4.66**	3.86 **6.70**
1000	1.08 **1.11**	1.13 **1.19**	1.19 **1.28**	1.26 **1.38**	1.30 **1.44**	1.36 **1.54**	1.41 **1.61**	1.47 **1.71**	1.53 **1.81**	1.58 **1.89**	1.65 **2.01**	1.70 **2.09**	1.76 **2.20**	1.80 **2.26**	1.84 **2.34**	1.89 **2.43**	1.95 **2.53**	2.02 **2.66**	2.10 **2.82**	2.22 **3.04**	2.38 **3.34**	2.61 **3.80**	3.00 **4.62**	3.85 **6.66**
	1.00 **1.00**	1.11 **1.15**	1.17 **1.25**	1.24 **1.36**	1.28 **1.41**	1.35 **1.52**	1.40 **1.59**	1.46 **1.69**	1.52 **1.79**	1.57 **1.87**	1.64 **1.99**	1.69 **2.07**	1.75 **2.18**	1.79 **2.24**	1.83 **2.32**	1.88 **2.41**	1.94 **2.51**	2.01 **2.64**	2.09 **2.80**	2.21 **3.02**	2.37 **3.32**	2.60 **3.78**	2.99 **4.60**	3.84 **6.64**

Answers to Odd-Numbered Exercises

Numerical answers depend upon the extent of rounding off the computations and upon the order of operations; consequently, the student should not expect to agree precisely with all of the following answers. Answers of a non-numerical nature are included to give the student a rough idea of the kind of answer expected for such exercises.

CHAPTER 2

5. Class boundaries are 0, 1, 2, 3, 4, 5, 6.

7. Skewed to the right because students with less than a C average will drop out or be removed from school and the mean is probably not much higher than halfway between a C and B average.

11. $\bar{x} = 4.5$.

13. (a) $\bar{x} = 45$, (b) $s = 4.65$.

15. $s = 2.1$.

17. 73% and 95%.

21. $\sigma \doteq 1$.

23. (a) The mean will be increased by 10 pounds but the standard deviation will not be affected. (b) The mean and standard deviation will both be increased by 10%.

25. $s = 2.01$; hence $(\bar{x} - s, \bar{x} + s) = (-2.01, 2.01)$, which includes all the data. Poor interpretation here.

27. Median $= 45$ and range $= 14$.

29. First quartile $= 3.1$ and third quartile $= 5.4$; hence interquartile range $= 2.3$. Ninth decile $= 7.2$.

CHAPTER 3

1. To each of the 8 outcomes for 3 tosses attach a fourth letter, H or T, to obtain the following 16 possible outcomes: HHHH, HHHT, HHTH, HHTT, HTHH, HTHT, THHH, THHT, HTTH, HTTT, THTH, THTT, TTHH, TTHT, TTTH, TTTT.

3. $\frac{1}{16}$.

5. (a) $\cdot B_2 B_1$ $\cdot WB_1$ (b) $\cdot BB$ $\cdot WB$
    $\cdot B_1 B_2$ $\cdot WB_2$ $\cdot BW$
    $\cdot B_1 W$ $\cdot B_2 W$

7. Frequencies depend upon the nature of the school and the amount of artificial hair coloring but there might well be at least twice as many brunettes as blondes and probably three times as many blondes as redheads. Expected frequencies would then be $\frac{6}{10}$, $\frac{3}{10}$, and $\frac{1}{10}$. Take a sample at the school to obtain estimates of the proportions.

9. (a) $\frac{4}{10} \quad \frac{3}{10} \quad \frac{3}{10}$
$\quad\quad \dot{e}_1 \quad \dot{e}_2 \quad \dot{e}_3$

11. (a) $\frac{1}{9}$, (b) $\frac{1}{6}$, (c) $\frac{1}{2}$.

13. (a) $\frac{11}{12}$, (b) $\frac{1}{4}$, (c) $\frac{25}{36}$, (d) $\frac{5}{6}$, (e) $\frac{4}{9}$.

15. (a) $\frac{4}{25}$, (b) $\frac{8}{25}$, (c) $\frac{4}{25}$.

17. The total number of points in A_1 and A_2 is given by $n(A_1) + n(A_2) - n(A_1 \text{ and } A_2)$ because $n(A_1) + n(A_2)$ counts the number of points in the intersection of A_1 and A_2 twice; hence

$$P\{A_1 \text{ or } A_2\} = \frac{n(A_1) + n(A_2) - n(A_1 \text{ and } A_2)}{n} = P\{A_1\} + P\{A_2\} - P\{A_1 \text{ and } A_2\}.$$

19. (a) $\frac{1}{6}$, (b) $\frac{16}{81}$, (c) $\frac{5}{18}$.

21. (a) .00763, (b) .00006, (c) .98461, (d) .01539, (e) .02313.

23. $P\{A_1 \text{ or } A_2\} = \frac{16}{36} + \frac{6}{36} - \frac{2}{36} = \frac{20}{36}$.

25. (a) $\frac{1}{10}$, (b) $\frac{3}{25}$, (c) $\frac{5}{6}$.

27. (a) $\binom{13}{4} \Big/ \binom{52}{4} = \frac{11}{4165}$, (b) $\frac{44}{4165}$, (c) $\binom{39}{4} \Big/ \binom{52}{4} = \frac{6,327}{20,825}$.

29. $p^2 + (1 - p)^2 = 1 - 2p + 2p^2$.

31. .0825.

33. (a) .44, (b) .70, (c) .052, (d) $.00\overbrace{\cdots}^{11}0157$.

35. (a) $\frac{4}{25}$, (b) $\frac{9}{25}$, (c) $\frac{16}{25}$, (d) $\frac{21}{25}$, (e) 1, (f) $\frac{12}{25}$, (g) $\frac{3}{4}$, (h) $\frac{2}{5}$.

CHAPTER 4

3. $P\{4\} = \frac{1}{9}$, $P\{5\} = \frac{2}{9}$, $P\{6\} = \frac{3}{9}$, $P\{7\} = \frac{2}{9}$, $P\{8\} = \frac{1}{9}$.

5. $\mu = 6$, $\sigma = 1.15$.

7. $\mu = 4.5$, $\sigma = 2.87$.

9. \$1.75.

11. $E(D) = 7$, $E(C) = 4$; hence choose dice (D).

13. $\$\frac{29}{16}$.

CHAPTER 5

3. Sixteen possible outcomes: AAAA, AAAN, AANA, etc.; hence $P\{0\} = (\frac{5}{6})^4$, $P\{1\} = 4(\frac{1}{6})(\frac{5}{6})^3$, $P\{2\} = 6(\frac{1}{6})^2(\frac{5}{6})^2$, $P\{3\} = 4(\frac{1}{6})^3(\frac{5}{6})$, $P\{4\} = (\frac{1}{6})^4$.

5. $P\{0\} = (\frac{3}{4})^6$, $P\{1\} = 6(\frac{1}{4})(\frac{3}{4})^5$, $P\{2\} = 15(\frac{1}{4})^2(\frac{3}{4})^4$, $P\{3\} = 20(\frac{1}{4})^3(\frac{3}{4})^3$, $P\{4\} = 15(\frac{1}{4})^4(\frac{3}{4})^2$, $P\{5\} = 6(\frac{1}{4})^5(\frac{3}{4})$, $P\{6\} = (\frac{1}{4})^6$.

7. More than 10 needed because $P\{x \geq 7\} = .88$. Eleven fuses will obviously make $P\{x \geq 7\} \geq .90$.

9. Trials are not independent. If it rains one day the chances are increased that it will rain the following day because storms often last more than one day.

11. $\mu = 2$, $\sigma = 1.15$.

15. (a) .16, (b) .66.

17. 60.4 inches.

19. (a) .01, (b) .07, (c) .76.

21. (a) $\frac{7}{64} = .11$ and .10, (b) $\frac{37}{256} = .14$ and .14.

23. .04.

25. .04, or .03 if geometrical refinement is used.

27. .13 as compared to .12, which is surprisingly close; however, one is for $x = 0$ and the other for $x \leq 0$.

29. (a) $P\{x\} = \dfrac{10!}{x!(10-x)!}\left(\dfrac{4}{10}\right)^x \left(\dfrac{6}{10}\right)^{10-x}$, (b) .45, (c) .46, (d) .90.

CHAPTER 6

1. Registrar's card files and random numbers, or, say, select twenty 10-o'clock classes at random and select five students at random from each class.

3. Individuals replying are usually kind-hearted or difficult-to-please customers. Most individuals will not bother to answer; therefore the sample will reflect the opinions of those with extreme views.

5. It would be necessary to discover whether the 600 who did not reply had the same views as those who did reply before one could trust the data obtained from the 400.

7. Households differ in size with respect to adults; therefore households with a large number of adults will not be adequately represented.

9. (a) .02, (b) .84, (c) .95, (d) .00.

11. The second curve should be three times as tall as the first curve at $x = 6$ and about one-third as wide.

13. .10.

15. $P\{z > 3.0\} = .001$; hence seem to be heavier, or the first hundred students to register may not be typical.

CHAPTER 7

1. (a) $P\{e < 5.88\} = .95$, (b) $P\{e < 2.94\} = .95$; hence twice the accuracy.

3. Using $z_0 = 1.96$, (a) 62, (b) 553.

5. Using $z_0 = 1.96$, (a) 62, (b) 385.

7. (a) $134.1 < \mu < 145.9$, (b) $135.1 < \mu < 144.9$.

9. (a) $P\{e < 1.7\} = .95$, (b) 138, (c) $40.3 < \mu < 43.7$.

11. .38.

15. $20.1 < \mu < 23.9$.

17. $27.9 < \mu < 33.2$.

19. $25.0 < \mu < 39.0$.

21. $P\{e < .055\} = .95$.

23. Using $z_0 = 1.96$, $n = 577$.

25. 601.

27. (a) Students first to arrive may not be typical. (b) $.23 < p < .37$.

CHAPTER 8

1. Assuming that an individual is innocent until proved guilty, the hypothesis to be tested is that an individual is innocent; hence the type I error is convicting an innocent individual and the type II error is letting a thief go free. Society considers a type I error more serious than a type II error.

3. $\alpha = \frac{1}{4}$, $\beta = \frac{3}{4}$.

5. $\alpha = .25$, $\beta = .91$; hence $x = 2$ is a much better critical region than $x = 0$.

7. $z = -1.33$; hence accept H_0.

9. $z = 3.3$; hence accept superiority.

11. $\bar{x} = 71.6$, $s = 31.4$; hence limits are 30–114. Control appears to exist, but data are too few to make reliable conclusions.

13. $t = -1.49$; hence accept H_0.

17. $z = -2.17$; hence reject H_0.

19. $z = -2.52$; hence reject H_0.

23. $t = .92$; hence accept H_0.

25. $t = 1.51$; hence accept H_0.

27. $z = 1.12$; no.

29. Honest mathematically means $p = \frac{1}{2}$, whereas honest from a practical point of view means that p is approximately equal to $\frac{1}{2}$.

31. $\hat{p} = .0255$; hence limits are .0105 and .0405. If percentages are used, multiply these answers by 100. Out of control on days numbered 18, 22, and 38.

33. $z = .75$; no.

35. $z = 1.33$; hence accept H_0.

CHAPTER 9

1. $r = .93$.

3. The formula shows that interchanging x and y has no effect on the value of r.

5. (a) .4, (b) $-.6$, (c) $-.5$.

7. If x denotes tide height and y denotes traffic density, then both x and y would tend to be large around 8:00 A.M. and small around 2:00 P.M.

11. 18, using one-sided critical region.

13. Since $|r| < .361$, accept H_0. No, since now critical value is $-.306$.

15. It should pass through the points $(-2, 2\frac{1}{2})$ and $(4, -\frac{1}{2})$.

17. $y' = 1.7 + 1.1x$.

19. $y' = 330 + .122x$.

25. $s_e = 27.6$, 25 per cent, 0 per cent.

27. $1.93 < \beta < 2.07$.

29. $t = .20$; hence accept H_0.

31. Choose x's as far apart as possible so as to make $\Sigma(x_i - \bar{x})^2$ as large as possible; hence choose five values at $x = 0$ and five values at $x = 10$.

33. (a) $y' = 3.38 + 2.36x_1$, (b) $s_e = 3.6$ as compared with 2.7; hence x_2 is beneficial.

35. (b) $r = .90$, (c) reject H_0, (d) $y' = 12.82 + 6.58x$, (e) $-4.4, 3.0, 2.7, 2.9, -12.7, -13.9, -3.9, -2.4, 6.0, 14.5, -4.3, 12.4$, (f) $s_e = 9.15$, (g) 0 per cent.

CHAPTER 10

1. $E(I) = 205$, $E(H) = 190$; hence choose ice cream concession.
3. $E(A) = 0$, $E(B) = 800$; hence choose B.
5. $E(100) = 400$, $E(200) = 500$, $E(300) = 100$; hence order 200.
7. $E(S) = 1000$; hence sponsor it.
9. (a) $P\{.4|S\} = \frac{3}{8}$, $P\{.5|S\} = \frac{3}{8}$, $P\{.6|S\} = \frac{2}{8}$ (b) $E(S) = -1250$; hence do not sponsor it.

CHAPTER 11

1. $\chi^2 = 2.4$, $\chi_0^2 = 6.0$; hence accept theory.
3. $\chi^2 = 11.1$, $\chi_0^2 = 6.0$; hence reject theory.
5. $\chi^2 = 4.5$, $\chi_0^2 = 12.6$; hence accept compatibility.
9. $\chi^2 = 4.7$, $\chi_0^2 = 5.0$; hence accept H_0. The decision is less positive than it was before.
11. $\chi^2 = 13.2$, $\chi_0^2 = 6.0$; hence reject independence.
13. $\chi^2 = 2.8$, $\chi_0^2 = 6.0$; hence accept same opinion.
15. $z = 2.2$; hence reject hypothesis. It can be shown that $z^2 = \chi^2$ value for this type of problem.

CHAPTER 12

1. $F = 13.6$, $\nu_1 = 3$, $\nu_2 = 20$, $F_0 = 3.1$; hence reject hypothesis that catalysts have no effect.
3. $F = 1.02$, $\nu_1 = 2$, $\nu_2 = 19$, $F_0 = 3.5$; hence accept no difference due to methods.
5. $F_c = 18.4$, $\nu_1 = 3$, $\nu_2 = 12$, $F_0 = 3.5$; hence accept that types differ. $F_r = 6.6$, $\nu_1 = 4$, $\nu_2 = 12$, $F_0 = 3.3$; hence accept that workmen differ.
7. $F = 10.8$, $\nu_1 = 3$, $\nu_2 = 12$, $F_0 = 3.5$; hence reject hypothesis of no plot differences.
9. (a) $F_c = 11$, $\nu_1 = 3$, $\nu_2 = 6$, $F_0 = 4.8$; hence reject no catalyst effect. (b) $F_t = 12$, $\nu_1 = 2$, $\nu_2 = 6$, $F_0 = 5.1$; hence reject no temperature effect.

CHAPTER 13

1. $z = -1.92$, $z_0 = -1.64$; hence reject H_0.

3. $t = 1.84$, $t_0 = 2.14$; hence accept H_0. The t value is quite close to the critical value as compared to the $z = 1.03$ value of problem 2, indicating that the t test is more sensitive to a shift in the mean than the sign test.

5. $z = 2.01$; hence reject H_0.

7. The pairs of values are not matched values here; therefore, there is no point in taking these particular differences.

9. $t = 4.10$, $t_0 = 2.10$; hence reject H_0. The t test appears to reject H_0 more easily than the rank-sum test does here.

11. $R = 153.5$, $z = -.71$; hence accept H_0.

Index

$P(A \cap B) = P(A \cup B) = P(A) + P(B) - P(A \text{ and } B)$

if A and B mutally exclusive $P(A \cup B) = P(A) + P(B)$

$P(A \mid B) = \dfrac{P(A \cap B)}{P(B)}$

$P(A \cap B) = P(A) P(B \mid A)$

$z = \dfrac{x - \mu}{\sigma}$

$\mu = \bar{x}$

$\sigma^2 = s^2$

regression line

$y' = \bar{y} + b(x - \bar{x})$

table 3

$r = \dfrac{(x_1 - \bar{x})(y_1 - \bar{y}) - (x_2 - \bar{x})(y_2 - \bar{y}) \ldots}{(n-1) \, s_x \, s_y}$ (stand dev.)

$b = \dfrac{\sum_{i=1}^{n} x_i y_i - n \bar{x} \bar{y}}{\sum_{i=1}^{n} x_i^2 - n \bar{x}^2}$

$b = \dfrac{(x - \bar{x}) y}{x - \bar{x}^2}$